GOLD IN QUARTZ

In 1918 a miner starts work on a stope at the 1300-foot level of the Old Brunswick mine, which later became a part of the Idaho Maryland mine.

GOLD IN QUARTZ

The Legendary Idaho Maryland Mine

JACK CLARK

COMSTOCK BONANZA PRESS
GRASS VALLEY, CALIFORNIA

NEVADA COUNTY PIONEERS SERIES

Published by Comstock Bonanza Press
18919 William Quirk Drive
Grass Valley, CA 95945-8611
Phone (530) 263-2906

ISBN 0-933994-31-1

This Book is Dedicated to

- The various managements who, for over a 106-year period, were able to leave a great heritage through their vision and perseverance.
- The hundreds of miners and other employees who worked so diligently to make this happen, and to those who paid the ultimate sacrifice.
- My wife and two daughters, who urged me to write this part of history for others to know about.

ILLUSTRATION CREDITS

All illustrations are from the author's collection except for the following: (CB = Charles Brown; CDMG = California State Division of Mines and Geology; DC = David Comstock; IMMC = Idaho Maryland Mining Corp.; JF = John Finn; MH = Malcolm Hammill; PHNC = 1895 Pictorial History of Nevada County; SHL = Searls Historical Library; TC = Thelma Collins; WG = William Goggin)

Frontispiece—SHL. **xii**—DC and Carleton Carter. **3**—(drawing) *Hutchings' California Magazine,* Oct. 1857; (photo) SHL. **4**—DC. **9, 12**—SHL. **15**—(top) Persis Coleman, in *Nevada County Narrow Gauge* by Gerald M. Best; (bottom) SHL. **17–18**—SHL. **24–25**—*History of Nevada County (1880).* **30**—Nevada County Historical Society *Bulletin* 33 (April 1979). **32**—DC. **35**—SHL. **36**—(top) PHNC; (bottom) SHL. **42**—SHL. **46**—DC and U.S. Geological Survey (1891). **49**—SHL. **50**—(top) Joan Jacks; (bottom) PHNC. **54–55**—SHL. **61**—*The Morning Union* (1895). **64**—PHNC. **66**—DC and Charles E. Uren. **68, 69, 71**—DC. **74**—SHL. **80**—SHL. **85**—DC. **89**—DC, IMMC and Charles E. Uren. **94**—Evelyn Foote Gardiner. **98**—MH. **103**—SHL. **116**—IMMC. **117**—(top) SHL; (bottom) DC. **122**—(top) SHL. **126**—SHL. **127**—(top) SHL; (bottom) DC. **135**—(top) JF; (bottom) CDMG. **138**—IMMC. **145**—(top) CDMG (W. O. Jenkins); (bottom) SHL. **150**—(bottom) IMMC. **152–153**—TC. **174**—WG. **179**—(top) TC. **180**—David Weisz Co. **182**—(top) IMMC; (bottom) TC. **183**—IMMC. **185**—DC. **191**—Tim O'Brien. **200**—DC. **204**—WG. **205**—MH. **211**—(top) Charles V. Litton Jr.; (bottom) DC. **215**—(bottom) IMMC. **226**—SHL. **232**—JF. **236**—DC and U.S. Geological Survey (1995). **245, 249**—CB. **251**—(top) CB; (bottom) DC.

CONTENTS

ILLUSTRATIONS AND MAPS

MAPS

PREFACE

During the 13 years that I worked for the Idaho Maryland Mines Corporation, my work gave me access to all of the surface and underground workings of the Idaho Maryland, Old Brunswick, New Brunswick, and Bullion mines. In that period I acquired considerable knowledge of the operations and underground geology. With this background and my interest in mining, I felt compelled to research the mine's history and record a factual account of this famous gold producer and the men who were an important part of it.

I began working at the Idaho Maryland mine on January 9, 194 31, as a mucker in a stope on the 1300-foot level. After two months I was elevated to First Aid Attendant in charge of safety and first-aid equipment on the surface and underground at the Idaho Maryland, Old Brunswick, New Brunswick, and Bullion mines. Between October 1942 and June 1945 the gold mines were closed by government order. During those war years I served in the South Pacific as a Pharmacist Mate in the United States Navy Reserve and Marine Corps. Following my Navy discharge I returned to the mine on November 3, 1945, and worked in the New Brunswick shop as a leadman, sharpening and tempering drill bits.

On May 1, 1949, I was elevated to the position of Safety Engineer. In addition to my safety responsibilities, I was given engineering assignments such as underground surveying, mapping stopes, and measuring miner's contracts. Then, for five months before I left the mine on November 30, 1956, I was the underground superintendent. The final mine closing occurred shortly thereafter.

The Idaho shaft did not follow in a straight plane, therefore there were several ups and downs. A series of metal rollers was attached to timbers along the footwall and hanging-wall timbers to protect the hoisting cable from wear and damage. The rollers were small in diameter and they turned at great speed when in contact with the cable, which traveled at 650 feet per minute. Several rollers could be in motion at the same time, and each gave off a different tone. Similar sounds could be heard in the No. 30 winze from the cable contacting the rollers. I traveled up and down the Idaho shaft hundreds of times, and I recall with pleasure the distinctive variety of sounds made by the rollers in the shaft.

JACK CLARK

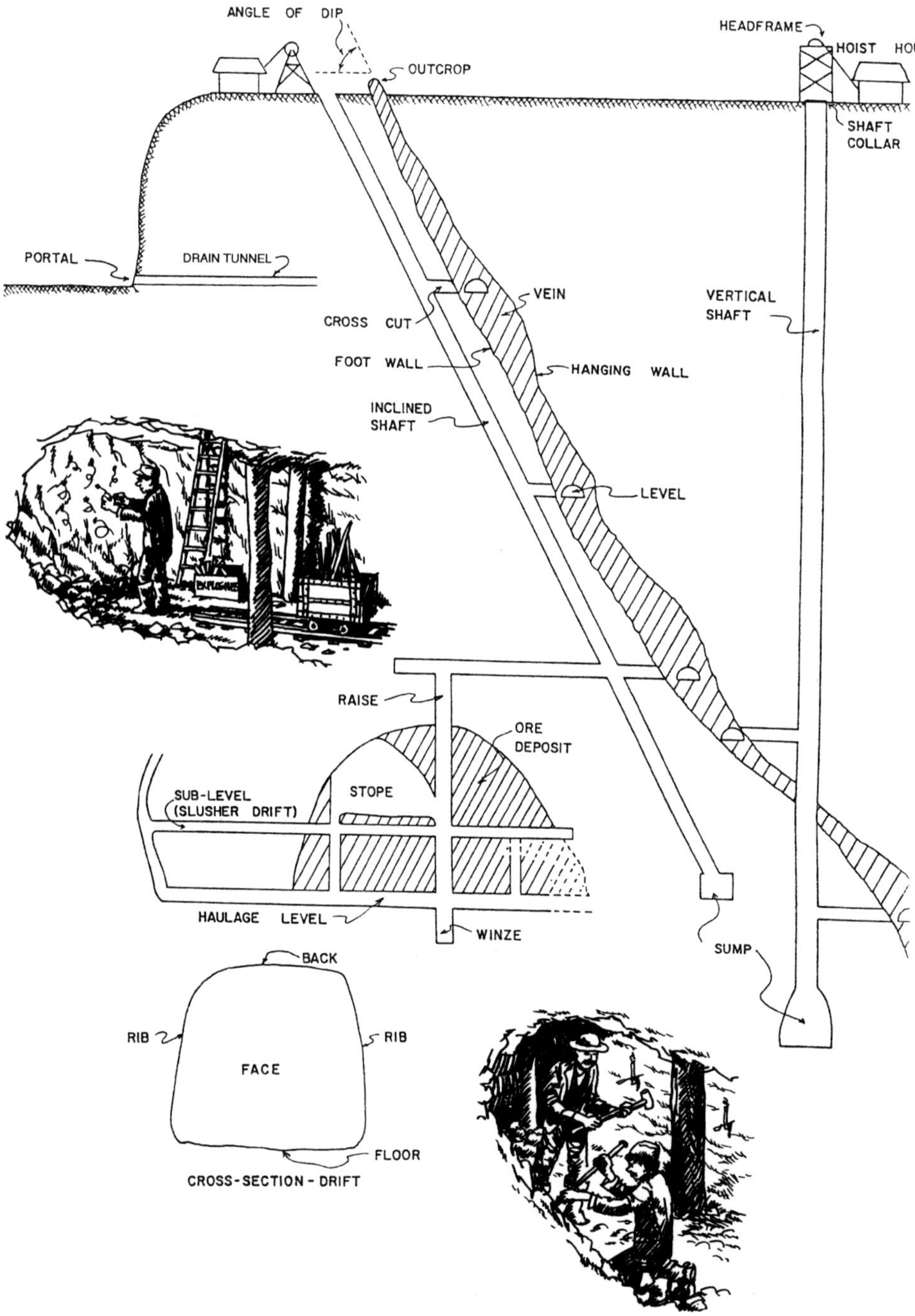
ANGLE OF DIP
OUTCROP
HEADFRAME
HOIST HOUSE
SHAFT COLLAR
PORTAL
DRAIN TUNNEL
VEIN
VERTICAL SHAFT
CROSS CUT
FOOT WALL
HANGING WALL
INCLINED SHAFT
EXPLOSIVES
LEVEL
RAISE
ORE DEPOSIT
STOPE
SUB-LEVEL (SLUSHER DRIFT)
HAULAGE LEVEL
WINZE
SUMP
BACK
RIB
RIB
FACE
FLOOR
CROSS-SECTION - DRIFT

GLOSSARY OF MINING TERMS

Adit. A horizontal passageway into a mine, driven from the side of a hill. Often referred to as a "tunnel."

Air Compressor. A machine that draws air in at atmospheric pressure, compresses it, and delivers it at a higher pressure.

Air Doors. Heavy wooden doors located at specific locations within underground workings to direct air circulation.

Air Drill. A percussion drill, or rotary-type rock drill, driven by compressed air, used principally to drill rock in underground workings.

Amalgam. An alloy of mercury with one or more other metals, primarily with gold or silver.

Amalgamation. The process by which mercury is alloyed with gold and silver from pulverized ores.

Apex. The highest point, or crest, of a mineral vein that reaches or is nearest the surface.

Assay. The testing of a sample of minerals or ore to determine the content of valuable minerals in the sample.

Back(s). The overhead portion of a vein in a drift, raise or stope.

Backfill. The placement of waste rock into worked-out stopes. Often referred to as "fills."

Bail. The cross-arm portion of the dumping mechanism on an ore skip where the hoisting cable attaches to the skip.

Ball Mill. A rotating type of gold recovery mill. Crushed ore is fed through the mill by means of a horizontal steel cylinder, and is pulverized by steel balls as the grinding media.

Barring Down. Removing loose rock from the overhead and sides of a work area, following a previous blast, by means of a long-handled bar. Also referred to as "scaling down."

Battery. The stamps that fall in one mortar; usually five stamps.

Beanhole. The rounds drilled upward from a drift to provide openings for chutes and to form pillars between chutes while starting a stope.

Bedrock. The solid rock of the earth's crust, generally covered by overburden of soil or water.

Blast. The detonation of explosives in drill holes, either singularly or in groups.

Bob. The engine beam on a Cornish pump to which the pump rod is attached.

Bonanza. A body of rich ore.

Bonnet. A metal covering over a mine cage to shield it from objects falling down the shaft.

Bootleg. The unbroken remains of a drill hole that had been blasted and failed to break cleanly.

Breast. The face of a working, usually in a drift or crosscut.

Buddle. An inclined shallow trough, used for separating ores in running water.

Bulkhead. 1) Heavy timbers covering raise sets to absorb the blast and support the weight of rock during raising operations. 2) A tight partition or stopping in a mine for protection against fire or gases.

Bulldoze. Blasting by securing a small quantity of explosives and detonator on or against a large rock by means of a soft mud.

Burncut. A pattern of holes drilled into the face of a drift or crosscut to provide the initial opening while blasting the round.

Bullion. Unrefined gold, melted and cast into a bar.

Cage. A fully enclosed conveyance within a vertical shaft for hoisting men and supplies.

Canvas Plant. A bank of long canvas bags placed in the tailings circuit to capture fine gold that otherwise would be lost.

Cap. The horizontal timber above the posts of a timber set.

Carbide Lamp. A lamp which operates by small drops of water on calcium carbide to form acetylene gas. When lit it produces an elongated flame. This lamp replaced the candle to provide greater light for working underground.

Chimney. An elongated body of ore standing vertically within a vein.

Chute. An inclined trough through which broken rock is transferred into mine cars from a stope or a raise.

Claim. A parcel of land legally held for mining purposes, the location of which is recorded.

Collar. 1) The term applied to the timber or concrete around the mouth of the shaft. 2) Also, the start of a drill hole; the point at which the drill steel first penetrates the rock face.

Color. Tiny particle of gold found within a mineralized zone.

Compartment. Raises are divided into two or more compartments to provide a separation of the manway and ore passage. Shafts and winzes are divided into two or more compartments to separate the hoisting from the manway.

Concentrates. The mineral portion of the ore which has been separated by concentration.

Concentration. The separation and accumulation of economic minerals from gangue.

Contract Miner. An unwritten agreement in which the miner is paid not for his time but for the lineal feet of advancement or cubic feet of rock broken.

Core Drill. A drill that penetrates a rock formation to remove a cylindrical core of rock for the purpose of examination and assay.

Cornish Pump. A single-acting engine in which the power for pumping operations is transmitted through the action of a cumbersome beam.

Counterbalance. Each skip or cage controlled by a double drum hoist can operate separately or together. Normally each unit is operated simultaneously from opposite elevations within the shaft or winze in order to equalize the weight.

Country Rock. The general mass of adjacent rock, as distinguished from a dike, vein, or lode.

Cousin Jack. A common nickname for a Cornish miner.

Cribbing. Close timbering, as in the lining of a shaft or timbered drift.

Crosscut. A horizontal passage driven across the course of a vein, or in general across the direction of the main workings. A connection from a shaft station to a vein.

Cyanide Process. A process for the extraction of gold from finely crushed ores, concentrates and tailings by means of cyanide of potassium. The gold is dissolved by the cyanide solution and subsequently deposited upon metallic zinc, or by other means.

Dead Work. Work not directly productive, though necessary for exploration and future production.

Depletion. The steady declining amount of ore in a deposit or property, resulting from production.

Development. Bringing a mining property to a production stage.

Dike. A fissure filled with solid material other than the one through which it cuts.

Diggers. Work clothes worn underground.

Dip. The downward inclination of a vein, with reference to a horizontal plane.

Dolly Hoist. A single-drum hoist used to transport men and materials into and out of a mine.

Double-Drum Hoist. A hoist consisting of two separate drums which, by means of a clutch mechanism, can operate one drum at a time or both simultaneously. Used mainly for hoisting rock.

Double Jacking. Drilling holes in rock manually, where one miner holds the steel and rotates it a fraction of a turn, while another miner strikes the steel simultaneously with an eight-pound sledge.

Downcast. A shaft through which air descends and is directed systematically throughout underground workings.

Drain Tunnel. A tunnel driven for effecting the general drainage of a mine, being designed to receive and carry off water both above and below its level.

Drift. A horizontal passage that is driven along the course of a vein, as distinguished from a crosscut, which intersects it.

Drive. To extend an excavation horizontally (a drift or crosscut), as distinguished from sinking or raising.

Dry. A building where miners change into their underground work clothes. The building is heated to allow the wet cloths to dry prior to the next wearing.

Dynamo. Early day term for a generator.

End Line. The boundary line, continuing vertically into the earth, between one mining claim and an adjacent claim or property.

Face. The end or portion or any underground working where drilling is to take place or has taken place.

Fault. A break in the continuity of a rock formation. It is accompanied by a movement on one side of the break or the other so that what were once parts of one continuous stratum or vein are now separated. The amount of displacement of the parts may range from a few inches to many feet.

Fissure. A long narrow opening, cleft, crevice or crack.

Flotation. The method of mineral separation in which a froth, created with reagents and water, floats some finely crushed minerals, while others sink.

Footwall. The wall or rock formation lying under a vein.

Gad. A small pointed iron tool used for cutting timber hitches and splitting rock.

Gangue. The undesirable minerals associated with valuable minerals in an ore deposit.

Gob. Miner's term for waste rock used to backfill worked-out stopes.

Gondola. A man-skip equipped with high metal sides and eight-inch-wide, ladder-like seats. This conveyance is designed for use in a steep inclined shaft to convey men into and out of a mine.

Goudge. A layer of soft clay or composed rock along the wall of a vein. Also referred to as selvage.

Grizzly. A grating of heavy rails spaced eight inches apart to prevent oversized rock from entering a transfer raise or shaft pocket.

Hanging Wall. The wall of rock or formation on the upper side of a vein.

Hardrock Mining. Mining in which drilling and blasting must be employed to break rock.

Haulage Level. Levels where ore or waste is trammed to shaft pockets for hoisting to the surface.

Headframe. The steel or timber structure at the collar of a shaft which carries the sheave wheels for the hoisting cables.

High-Grade. 1) Rich ore in which an abundant amount of gold is showing. 2) To steal gold from a mine.

High-Grading. The theft of specimen gold by mine employees.

High-Grader. One who steals and sells or otherwise disposes of high-grade or specimen gold.

Hitch. An indent or hole cut into the footwall or hanging wall in which to anchor a timber.

Hoist. A machine which raises and lowers the cage, mantrucks and skips in a shaft or winze.

Horse. A mass of country rock lying within a vein.

Incline. Any shaft, winze or other workings that follows the dip of the vein, either up or down.

Intermediate Level. A level driven from a raise and not yet connected to a shaft station.

Jim Crow. A tool used to bend mine rails to form the turns in a drift.

Lagging. Wooden planks used with raise timber to separate the manway from the ore passage, and also above or on the sides of drift timbers where necessary to prevent loose rock from falling.

Leasing. A variation of contract mining in which a group of miners lease a mine or, more often, sections of a mine.

Ledge. Synonymous with a vein.

Level. Underground mines are customarily worked from shafts through horizontal passageways, namely drifts or crosscuts. These are spaced at regular intervals in depth and are either numbered from the surface in regular order or designated by actual elevation below the collar of the shaft.

Lode. A clearly defined vein of rich ore separated from the adjoining rock by definite boundaries.

Magazine. A section of unused drift or crosscut, away from active workings, that is partitioned off by a metal-clad wall and door, where a limited supply of stick dynamite is stored.

Mantruck. A truck provided with seats and hand holds in which men are transported in a winze or inclined shaft. Often referred to as man-skips.

Manway. A separate timbered section of a raise, winze or shaft, providing access and egress by means of ladders or steps, depending upon the degree of dip.

Marginal Ore. Lower grade ore which is close to being uneconomical to mine.

Mine Dump. A pile of waste rock located on the surface of a mine operation.

Miner's Inch. The quantity of water that will escape from an aperture one inch square through a two-inch plank, with a steady flow of water standing six inches above the top of the escape aperture.

Missed Hole. A drill hole containing an explosive charge that failed to explode.

Mercury. A heavy silver-white metallic chemical element, liquid at ordinary temperatures, which sometimes occurs in a free state but usually in combination with sulfur. (Quicksilver).

Mortar. A tublike receptacle with grated sides, in which ore is stamped.

Muck. Broken rock, either ore or waste.

Mucker. One who loads broken rock into ore cars by hand or by machine.

Muck Pile. The broken rock resulting from blasting a round of drill holes.

Moil. A short pointed steel used to cut hitches for placement of timber.

Nipper. A worker assigned to distribute tools and supplies within a mine.

Ore. A mineral of sufficient value as to quality and quantity that it may be mined with profit.

Ore Reserves. The tonnage of ore of a certain grade which is estimated for the mine or a certain section of it.

Oreshoot. A large and usually rich aggregation of mineral in a vein.

Outcrop. The part of a stratum or a vein that appears at the surface of the ground.

Pillar. An undisturbed block of rock left during stoping operations to provide hanging wall support where deemed necessary.

Pinch. A vein when it narrows in width.

Pipe Crow. A tool used to bend pipe underground.

Pitch. The dip or inclination of a vein or an ore body in the direction of its strike.

Pocket. A large excavation for storage of rock, adjacent to a shaft or winze, in which mined rock is dumped from ore cars prior to being hoisted to the surface.

Portal. A surface opening of an adit or tunnel.

Pull. A term relating to the length of advancement of a round through drilling and blasting.

Quartz. A crystalline mineral, often transparent, in which gold and silver veins are most commonly found.

Quicksilver. A common name for mercury.

Raise. A vertical or inclined excavation driven upward from a level to explore the vein and provide access for stoping operations.

Rake. The direction of an ore shoot within a vein.

Safety Dogs. All cages in a vertical shaft are equipped with spring-loaded catches, or "safety dogs," designed to automatically stop and hold the cage at any point in the shaft, in the event the hoisting rope should break.

Safety Fuse. An internal burning medium through which fire is conveyed at a continuous and uniform rate for the purpose of firing explosives.

Sample. A portion of ore systematically taken, by which its quality is to be analyzed.

Set of Timber. The timbers which comprise any framing, whether used in a shaft, drift, winze or raise.

Set Up. To place a drilling machine, air hose and water hose in position for drilling.

Shaft. A passageway leading from the surface into underground workings, through which mining operations are conducted. A shaft may be "vertical" or "inclined."

Sheave Wheel. A large diameter wheel with grooved rim which guides the hoisting rope into the shaft, and is located on top of the headframe.

Shift. The length of time a miner works in one day; the regular work period for all miners, such as the day shift or night shift.

Shift Boss. A man who is in direct supervision of men working in a specific section of a mine.

Shrinkage Stope. A method of stoping which utilizes part of the broken ore to remain in the stope for the miners to work off of to drill subsequent rounds.

Sink. To excavate downward, such as in a shaft or winze, by drilling and blasting.

Skip. A self-dumping conveyance for hoisting ore or waste rock, which slides between guides in a vertical shaft, or on wheels and runs on tracks in an incline shaft or winze. The bail is attached to the lower portion of the skip in order that it may dump automatically.

Skip Tender. One who controls operation in a shaft or winze by loading and unloading men and supplies and loading rock for hoisting.

Slide. Two 1" by 12" boards positioned side by side along the footwall timber of a raise, in which a slide bucket operates to hoist drills, steel and timber for raising and stoping operations.

Slime. Ore reduced to a very fine powder and held in suspension in water so as to form a kind of thin mud.

Slusher. A mechanical drag shovel used to move ore or waste underground.

Spit. The end of safety fuse is slit open to expose the powder train prior to ignition. The powder gives a spitting sound when lit; thus the term "spit" evolved.

Spitter. A hot-wire igniter for lighting blasting fuse.

Square Set. A specific set of timbers used for support in underground mining.

Station. An excavation adjacent to a shaft or winze on each of the levels to provide sufficient room to load and unload men and materials.

Stamp Mill. A mill in which the ore is pulverized to 30-40 mesh by means of heavy stamps.

Steel. Steel drill rod for drilling holes for explosive charges.

Stench. A warning system by which methyl mercaptin can be injected into the underground compressed-air system to immediately alert underground personnel, in the event of a fire or other dire emergency.

Stope. An underground excavation along a vein from which ore is being, or has been, extracted, as distinguished from other excavations such as drifts, raises or winzes.

Stringer. A narrow vein of irregular filament of mineral traversing a rock mass of different materials.

Strike. The course or bearing of the outcrop of an inclined bed or structure on a level surface; it is perpendicular to the direction of the dip.

Stull. A timber used to support a section of hanging wall or other rock to render the area safe to work under.

Sub-Drift. An exploratory drift, usually driven from a stope or raise.

Sub-Level. An intermediate level opened a short distance from a main level.

Sump. An underground reservoir where water collects and from which it is pumped.

Tails. The discarded sands from the milling process after the gold has been extracted.

Tailings. The waste material resulting from the milling process.

Tamp. Confining the explosives solidly in a drill hole with stemming, to guard against accidents as well as to secure the best execution from the explosive. A long wooden rod is used to prevent sparking.

Throw. The displacement of a vein and surrounding rock formation caused by a fault.

Timber. The wooden supports and braces used to stabilize underground workings.

Tram. To push ore cars in a mine.

Tributers. Miners who agree to accept for their labor a portion of the proceeds of the ore taken out.

Tugger. A small air-operated hoist used to unload timber at shaft stations, and for hoisting supplies in a raise or stope.

Upcast. A shaft through which air that has circulated through underground workings ascends and is exhausted to the surface.

Vanner. A machine for dressing ore; an ore separator; a vanning machine.

V-Cut. A cut where the initial rock blast out of the face is like the letter V. It usually consists of six or eight holes drilled into the face, half of which form an acute angle with the other half.

Vein. A zone or belt of mineralized rock within boundaries clearly separating it from neighboring wall rock.

Vertical Shaft. In a vertical shaft men ride in a cage which operates between wooden guides, and is equipped with dogs or catches of standard design to hold the fully loaded cage should the hoisting rope break.

Waste. Rock that is too low in grade for economic value.

Winze. A vertical or inclined passage that starts from an underground level and usually connects with one or more lower levels.

Workings. The various underground locations, such as drifts, crosscuts, stopes, winzes, etc., where work is progressing or has taken place.

INTRODUCTION

The whispered word "gold" has acted on humans like a fever ever since they first learned to cherish this glittering metal. In California no other event in its history has equaled the explosive influence of the gold rush. A remote area in the western part of North America suddenly became the center of worldwide attention. Men of all nationalities focussed their dreams on obtaining instant wealth.

The California gold rush began on January 24, 1848, on the American River at Coloma in the Sierra Nevada foothills. There James Marshall found several gold nuggets in the tailrace of a sawmill he was constructing. His discovery occurred within months of the transfer of California to the United States from Mexico.

Men and women came from nations worldwide in search of the precious metal. They traveled by wagon, on horseback, on foot, and by ship. Seekers endured great hardships as they struggled to reach and stake out their claims. New territories were explored and developed; towns grew up overnight.

Within a short time following the discovery of gold at Coloma, placer gold was also found in Wolf Creek, Deer Creek, and the Bear and Yuba rivers. In 1849 and 1850 many miners were working the extremely rich but comparatively shallow placer deposits of this district. The rapid exhaustion of placer gold turned the miners' attention to gold-bearing quartz veins.

In October 1850 George McKnight discovered gold in a quartz vein on what later became known as Gold Hill in Grass Valley. The vein was quite rich where it outcropped on the surface, and the miners' attention quickly turned to locating other gold-bearing quartz ledges in and around Grass Valley. The exceptional richness of ore from many of the newly located quartz veins caused lode mining to be more rapidly developed in the Grass Valley mining district than any other district in California.

From the exceptional richness of ore in the many newly discovered quartz veins grew the mighty Empire, North Star, Brunswick and Idaho Maryland mines. Underground hardrock mining went on to sustain the local communities for the next 106 years.

CHAPTER 1: EUREKA MINE

The Eureka, Idaho and Maryland mines were originally separate mines with one vein in common. This vein was marked by a conspicuous outcrop of white quartz on the Eureka claim. It was less pronounced on the surface as it entered the Idaho ground, and could be traced on the surface for only a short distance east of the Idaho shaft. A rich ore shoot within the Eureka mine was later found to extend into each of the other two mines, and thus became known as the "Eureka-Idaho-Maryland ore shoot."

Had it not been for the Eureka mine and the discovery of this remarkable ore shoot, the Idaho and Maryland mines likely would not have been developed. Although the Eureka mine did not become a part of the Idaho Maryland until 1919, it seems logical to begin this history with the Eureka mine.

This mine was situated on the north side of Wolf Creek on what was known as Eureka Hill, about one and one-quarter miles east of Grass Valley. It was located on February 7, 1851, and was one of the earliest quartz locations in this district. The ledge was recorded in rectangular claims, which measured 30 by 40 feet, and extended for 1800 feet. The mine was first worked at intervals by Bertrand L. Lamarque, George D. Roberts and William Chollar. The quartz close to the surface was low grade and could not be mined at a profit, so in 1857 Roberts and Chollar sold their interest to Lamarque; he in turn sold his holdings to Jules Fricot, Sidouin Ripert and Alphonse Pralus that same year.

In 1858 five thousand tons of quartz were taken from above the drain tunnel level, which was 30 feet below the surface. The yield was less than ten dollars per ton and did not pay expenses. Commencing in 1863 a perpendicular shaft was sunk to a depth of 100 feet, where a short crosscut exposed the vein. Drifts and stopes were started on this level as soon as possible. The shaft continued downward another hundred feet during 1864. In both levels the vein was found to be large and exceedingly rich, increasing in width and quality of ore as the vein was worked downward. At 50 feet from the surface the quartz ran $15 per ton, and increased to $28 at 100 feet, and averaged $37 between the 100- and 200-foot levels. The Eureka vein ran nearly east and west, and dipped to the south at an angle of 78°, averaging three feet in width.

Seventy men were employed on the surface and underground. Mark Bennallack was the underground superintendent. Sufficient gold was removed from these two levels to pay for the installation of a 20-stamp mill, hoisting works, and pumping unit, all at the expense of $60,000, as well as paying good dividends to the three partners. Fricot and company lacked the capital to develop the mine properly. On October 1, 1865, the property was sold to a group of investors for $400,000, at which time the Eureka Gold Mining Co. was incorporated. The new owners were J. B. Dickerson, Thomas Hope and Benjamin Silliman of New York; Edwin Hull, W. H. Cronise, John C. Winans, Milton Buckly, James Freeborn, A. J. Hope, Robert F. Morrow, N. J. Hall, George W. Beaver, L. S. Adams and Francis Berton of San Francisco; William Watt, Robert Watt, Jules Fricot, Alphonse Pralus and Sidouin Ripert of Grass Valley.

The Eureka Gold Mining Co. took over control of the mine on October 1, 1865, and under the management of superintendent William Watt, the mine began producing ore right away. The vein on the 100- and 200-foot levels was three to four feet in width, and drifting and stoping were accelerated in very good ore. An incline winze was started, and soon the 300-foot level was opened for exploration. The work force was steadily increased to 120 men, including both surface and underground operations. During the first year of operation the new company milled 11,460 tons of ore with a yield of $50 per ton. This gave a total value of $531,431, of which $380,000 was paid in the form of dividends to the 20 partners.

In the fall of 1866 a four-compartment shaft was started 170 feet west of the original shaft. This incline measured 5 by 20 feet, and was

Artist's conception of the underground workings of a California gold mine in 1857.

William Watt of Grass Valley.

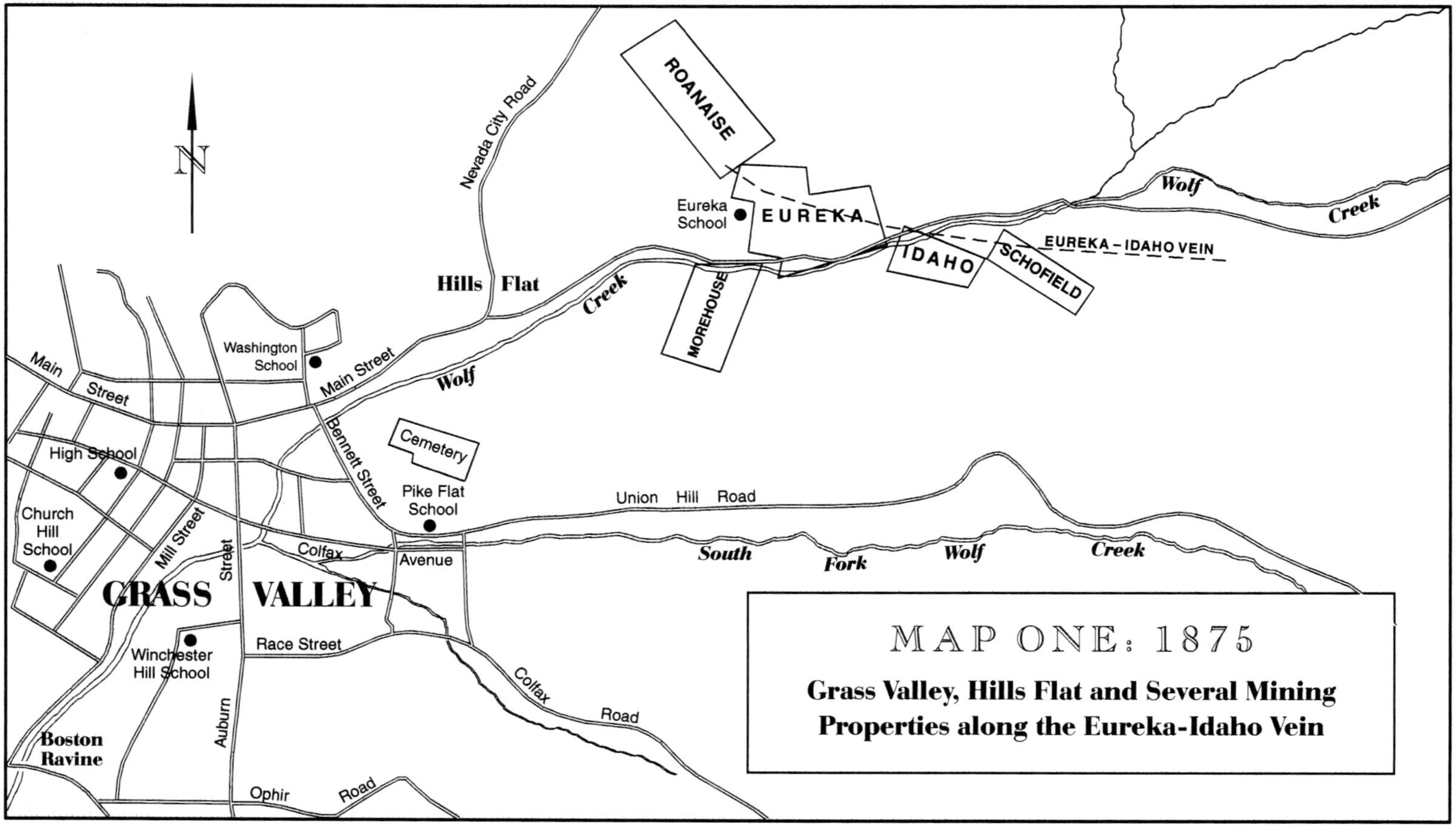

MAP ONE: 1875

Grass Valley, Hills Flat and Several Mining Properties along the Eureka-Idaho Vein

designed to develop the vein in depth. There were two compartments for hoisting ore and men, one for shaft sinking, and one for the Cornish pump rods. Each compartment had a hoist powered by steam. The hoists had flat steel ropes with winding wheels. The safety cages operated on wooden guides, and were designed to hold one tram wagon at a time. The shaft was raised from the 300-foot level while being sunk from the surface at the same time.

Twice within the first two years of operation the mill could not keep up with production, and ore was hauled to the Sebastopol mill in Boston Ravine to be milled. This caused additional expense, as in those days horses and wagons were the only available means of hauling rock. Thanks to the richness of the Eureka ore shoot, the mine quickly took its place as the leading gold quartz mine in California. Some even reported the Eureka to be the leading gold-producing mine in the world.

Eureka Stamp Mill

The mill was built in 1864 and contained twenty stamps, each weighing 850 pounds and dropping 60 times a minute from ten inches. Ten stamps were added in 1867, each weighing 700 pounds and dropping 68 times a minute from ten inches. The ore that was fed to the batteries was crushed to two inches in a Varney rock breaker. Water was fed into each mortar box, forcing the pulverized quartz through a 40-mesh brass wire screen, then over copper aprons covered with wool or cotton blankets. The blankets were constantly washed out into tanks to collect the free gold and sulphurets caught by the blankets. The blankets were found to catch 90 percent of all gold saved by the mill.

The blanket sands were treated in three Atwood amalgamators, which consisted of level troughs with quicksilver at the bottom; and over the troughs were horizontal revolving cylinders with projecting spikes which stirred up the quicksilver and pulp as the latter passed over the trough. The skimmings from these were ground in two Knox pans. The sands from the amalgamators were passed over Hunter Eureka rubbers, a table provided with oscillating rubbers suspended over an amalgamated copper surface with mercurial riffles, and designed to catch particles of amalgam that might otherwise escape. The labor required to run the mill for each 24 hours was: four rock breakers, four feeders of stamps, four blanket washers, two engineers and one amalgamator.

The sulphurets were processed by a chlorination process under the supervision of G. Fred Deetkin. In general this chlorination process separated the gold from the concentrates by first roasting the sulphurets. They were placed in an oven and brought to a red heat and retained in that condition for six hours, or until the smell of sulphur disappeared. The sulphurets were sprinkled with water after they had cooled, and then placed into wooden boxes, so constructed that chlorine gas could be introduced at the bottom and made to rise all through the mass. The boxes were kept closely covered, and chloride of gold, which is soluble in water, was formed. Water was introduced after four or five hours and chloride of gold was dissolved by the water. The solution was then drawn off into glass vessels in which iron sheets were placed. The chlorine united with the iron, and the gold fell out as a purplish brown powder to the bottom of the vessel.

Underground Operations—1868-1869

In 1868 underground operations continued in excellent ore, and new ground was opened by drifts driven east and west on both the 300- and 400-foot levels. The vein maintained its width, ranging from three to four feet, with good mineralization. Sulphurets were about 1½ percent of the quartz, containing mainly iron pyrites, and some chalcopyrite, lead and zinc. The footwall consisted of a soft serpentine, and the hanging wall a diabase dike. From October 1, 1865, until September 31, 1868, there were 33,319 tons of ore milled, with a return of $1,558,701. During that period $670,000 had been paid out in the form of dividends.

Underground foreman Mark Bennallack fell down the Eureka shaft a distance of 200 feet and was killed instantly. The accident occurred on March 20, 1868. His brother Frank Bennallack was killed by a cave-in at the Eureka mine about a year earlier.

The Eureka vein dipped 78°, therefore the stopes and raises were very steep. Often the rock would wedge together so as to form a bridge. A large void would then occur when the ore was pulled from the chutes below. A miner who attempted to dislodge the rock from underneath exposed himself to severe injury. The company then secured a small artillery piece, which weighed about 250 pounds. The small cannon was taken underground and would be aimed at the bridged area through a chute opening. The black powder was ignited remotely and an iron projectile was fired into the obstruction, causing the broken rock to fall. The projectile was retrieved and reused. This operation proved very satisfactory.

The new incline shaft continued in its downward advance, and by September 31, 1869, the shaft was down 60 feet below the 500-foot level. The shaft followed the vein at 78° down to the 300-foot level, and from there the shaft and vein flattened to 66°. There were 21,526 tons of ore hoisted during the year, of which 125 tons were crushed at the Sebastopol mill and the remainder at the company's mill, averaging 66½ tons per day with 30 stamps operating.

The 400-foot level had been driven a total of 1,418 feet in length, and the 500-foot level had been extended 640 feet. The 400-foot level had already produced 37,485 tons of quartz and 1,307 tons of ore had been removed from the 500-foot level. The total proceeds for the year amounted to $574,964, with a profit of $361,211, from which $264,000 was paid in dividends.

The Eureka company obtained a U.S. patent on 23½ acres of property covering the Eureka lode for a length of 1,680 feet. In addition, the company purchased the Roanaise mine that adjoined the Eureka on the west. It comprised 2,000 feet in length on the lode, giving the Eureka company a total length of 3,680 feet along the lode.

Miners Strike—1869

Every foot of underground advancement required drilling and blasting, but the tools with which miners worked in the early days were primitive. Drilling rock at that time was accomplished by one miner holding and turning the steel, while another miner struck the end of the drill rod with an eight-pound sledge hammer. Black powder was used to blast the holes.

In 1868 a new blasting agent called "giant powder" was introduced in the Grass Valley mines. This nitroglycerin-based blasting powder in stick form produced effects vastly superior to those resulting from the use of black powder. Unlike the latter, this powder would seldom blow out of the drill hole, but rather break away the bottom and sides of the hole in all directions, giving good fragmentation of the rock.

Giant powder revolutionized drilling operations because of its greater breaking strength. Large diameter drill holes and two-handed drilling were eliminated, and smaller drill steel was introduced. One miner could drill as many holes with the smaller steel as two men with the larger ones. This was a great savings in manpower, lowering the overall mining costs significantly.

Each miner drilled separately, using a four-pound hammer, and when he was finished in one location, he would be sent to another to drill. A few trusted men on each shift were assigned to load and blast the drill holes, then remove the broken rock on the subsequent shift. This change was made to limit access to specimen rock in order to reduce the temptation to high-grade any gold.

On May 10, 1869, the Miners' League called a strike at the Empire mine. The purpose of the strike was to prevent the continued use of giant powder in all mines of the district. The union claimed giant powder caused headaches and was injurious to the miner's health. In addition, the Miners' League asked that the pay for surface workers, muckers and trammers be increased from $2.50 to $3.00 per day.

The strike was ill advised, for giant powder had been used to a great advantage for over eight months before the Miners' League decided to strike. Although health reasons were cited, it was generally accepted that the introduction of single-hand drilling and the exclusion of miners from specimen rock were the underlying reasons for the action.

The Empire mine was forced to suspend operations when the League members refused to report for work and threatened new miners hired to take their places. After four weeks, Captain S. W. Lee, superintendent of the Empire mine and mill, announced that the mine would reopen using black powder for blasting in place of giant powder. A large group of miners, consisting of new men along with regular miners, reported to work. The union felt that it had conquered the giant powder issue, but the strike and threats continued. The League contended that only the original work force should be rehired. The strike was very unpopular with the business people due to their revenue losses.

The strike did not spread to the Eureka and Idaho mines, even though giant powder was in use. Within two months the strike ended, after the miners began returning to work on their own. It was only a short while until single-hand drilling and giant powder were used in all mines.

Underground—1870

The Eureka mine reached its maximum production in 1870 with the ore coming from the 300-, 400- and 500-foot levels, which were the heart of the Eureka ore shoot. The mill was running all 30 stamps on a 24 hour basis. There were 20,562 tons of ore milled, valued at $661,893, a yield of $31.75 per ton. A total of $400,000 was paid in dividends during 1870, the largest amount in any 12 month period.

By September 1871 the main shaft had reached a depth of 849 feet on the incline, or 786 feet vertically below the collar. Four compartments were carried to the 700-foot level, where the size reduced to just two compartments. The diabase contact was followed, but at the 700-foot level the contact dipped 48° in a southerly direction.

The shaft continued in rich ore to just below the 600-foot level. From this point to the bottom of the shaft, the vein maintained an average of 2 to 3 feet, but the tenor of the ore averaged only $7 to $8 per ton, as compared with a value of from $30 to $60 per ton in the main ore shoot.

Miner lighting fuse with a candle.

There were two distinct veins in the Eureka mine which were separated from one another by the diabase dike. The hanging wall vein was narrow and very weak on each of the levels where it was cut by crosscuts. The size and grade of quartz did not warrant exploration.

The mine continued to produce well during 1871, although the amount of ore milled was less than that of the previous year. The ore hoisted came from the 300-, 400- and 500-foot levels, but mostly from the latter.

Miners Union

The miners union called a strike against the Empire mine back in 1869 that lasted nearly two months. They claimed the use of giant powder was injurious to the miners' health. Those who wanted to work were threatened with injury if they crossed the picket line.

Then on Febraury 20, 1872, the miners union began threatening miners at the Eureka mine. Again the union did not want miners to use giant powder. Two miners were walking home after the night shift when someone shot at the men with a shotgun. B. F. Steese was struck with pellets on his chin, left arm and leg. In another incident David Rogers received a serious head injury when he was struck behind the ear with the butt end of a billiard cue.

A strike was called against the Eureka mine and the threats continued for anyone wanting to work. Then on March 15, 1872, the miners union, under pressure from the general public, rescinded its former resolutions against the use of giant powder. The miners were free to return to work.

In Search of New Ore—1873

The mine had flourished for several years under the able management of superintendent William Watt, but now the company had come to the crossroads, where a new ore body had to be found. The ore blocked out on the 500- and 600-foot levels was only enough to carry through perhaps another two years.

In 1873 the west drift on the 500-foot level came to a split in the vein. The large split, and what seemed to be the main vein, curved toward and pitched southerly under Wolf Creek. This split was followed, but the quartz was comparatively barren. The narrower split bore slightly northward from the true course of the ledge, and this, too, was drifted upon, but again was low grade. The 500-foot level had been in good ore until the split in the vein occurred.

In July 1873 a prospecting shaft measuring four by eight feet was started on the surface, about 200 feet back from the Eureka west boundary line. A well defined quartz vein that measured two feet in width, including the goudge, was followed. The dip and general characteristics of the quartz appeared to be the same as that of the north split found on the 500-foot level. The prospect shaft was sunk to a depth of 105 feet before it was stopped. Nothing of value was discovered. A shaft was also sunk 53 feet in depth on the Roanaise claim in search of a new ore body, but again it was disappointing.

During the years 1872–1874, the main thrust of development was to find a new ore shoot to extend the life of the mine. The shaft-sinking operation was stopped at 1,250 feet, because the grade of quartz was too low to warrant any further development.

During these three years the ore production dropped significantly. The additional distance and time to hoist rock from the lower levels reduced the amount of ore produced daily. The reduction of ore resulted in cutting back the milling operations. The mill had been operating 30 stamps and running 24 hours a day up until the first part of 1871. From that point on the stamps in use were reduced to 10 to 15 on a daily basis.

The rake of the Eureka ore shoot was nearly horizontal, beginning at a depth of 100 feet and ending at a depth of 600 to 700 feet. The ore shoot was worked for a length of approximately 1,600 feet.

Additional properties were acquired in hopes of discovering another ore body. The adjacent Whiting ground was acquired and the Mobile and Roanaise claims were purchased. The price, plus perfecting titles, amounted to $302,706.

At a distance of 1,000 feet west of the Eureka shaft the vein was lost, and considerable exploratory work, including crosscutting in the Roanaise and Morehouse claims, failed to find the westward continuation of the ore shoot. This fact, in conjunction with the poor showing of the mine below the 600-foot level, caused a suspension of work on June 23, 1877.

The Eureka mine was the leading gold-producing quartz mine in the United States for a period of six consecutive years, 1866 through 1871. The total production of the Eureka mine is estimated to have been $5,700,000, with dividends in the amount of $2,094,000.

This 1877 photo of prominent Grass Valley pioneers includes (39) Alfred B. Dibble, (22) Alex. B. Brady, (36) William Watt, (31) John Coleman, (29) Edward Coleman and (41) Peter Johnston.

CHAPTER 2: IDAHO MINE – PART I

The Idaho mine was situated about a mile and a half east of downtown Grass Valley. The ledge was first discovered in the bed of Wolf Creek during an early period of placer mining. The claim was not recorded until May 9, 1863, after the rich strike was made in the adjoining Eureka mine. The claim filing included the names of Thomas J. Pegg, Alfred B. Dibble, William S. Byrne, Henry D. Cady, James B. Mason, George D. McLean, Alexander B. Brady, S. D. Leavitte, Joseph O'Keefe, P. Lyda, C. Pralus, Myer Cohen, Alphonse Pralus, C. Chabreaux, George W. Dixon, Edwin W. Maslin, F. Braumburt, James K. Byrne, William Young, William Loutzenheiser, Thomas Findley, Peter Johnston, C. C. Clark, E. Fellers, John Weber, Jacob Moris, Miles P. O'Connor, Jules Fricot, Sidouin Ripert, Charles Verdelais and Frank G. Beatty.

Little attention was given to working this claim at first, other than the yearly assessment work necessary to hold the claim. The quartz taken from the vein during this work proved to be quite low grade. The vein was narrow, therefore no effort was made to proceed any further. Early in 1865 work commenced in a systematic manner with the intent of following the outcropping as far as possible. This proved very disappointing, as the vein could not be traced for more than 175 feet.

In the spring of 1865 a prospecting hole was sunk close to the north bank of Wolf Creek, where the quartz improved enough in character that the company felt justified in commencing work on a large scale. Then a 12" pumping unit and an 8" hoisting engine were set in place. On July 5,

1865, a perpendicular shaft, 4½ by 9 feet in size was started on the south side of Wolf Creek, about 290 feet from the western boundary line.

This shaft was located so as to intersect the ledge 150 feet from the surface. It was sunk to a depth of 120 feet, where a crosscut was driven into the footwall. A small vein was struck 8 feet from the shaft, but as it was only a few inches in size and showed no gold or mineral, the vein was passed by and the crosscut continued another 50 feet. The small vein later proved to be the main Eureka-Idaho vein.

Up until this time the copartnership company had expended $19,496.50. Some of the members were unable to pay their share of expenses, so it was decided to close the work until sufficient capital could be obtained. In March 1866 all work was stopped, and the mine remained closed for about 18 months.

New Company Formed—1867

In September 1867 John and Edward Coleman, having just sold the North Star mine, bought out the original group of owners. They organized the Idaho Quartz Mining Co. with a capital stock of $310,000, divided into 3,100 shares, each representing one foot on the ledge. The elected officers were John C. Coleman, president; William Young, vice president; Miles P. O'Connor, secretary; Thomas Findley, treasurer; and Edward Coleman, superintendent. Each officer, with the exception of the Coleman brothers, was from the original group.

Work commenced again in sinking the shaft. At about 140 feet from the surface the fissure was struck, but there was no quartz present. There were only a few inches of clay matter where the ledge should have been. The shaft was continued to 300 feet from the surface without meeting any encouragement. At this point a drift was driven west, and at 108 feet from the shaft the ledge widened, showing considerable gold. The first 20 tons of ore yielded $29 per ton.

After a few months of successful exploration a decision was made to erect a stamp mill, which was completed in October 1868. There were 763 tons of ore milled that same year, yielding $13,978 from the newly completed mill, plus $31,557 from outside mills. The expenditures for the first 15 months (which included all equipment, buildings and mining operations) amounted to $69,098. The full amount of money paid by the stockholders for the development of the mine since incorporation amounted to $38,550. This was the last time assessments to stockholders were necessary.

John C. Coleman.

Edward Coleman.

Early underground scene in the Idaho mine.

After striking the ore shoot on the 300-foot level the shaft was continued downward and levels were established at 100-foot intervals. The quartz widened to two feet or more, with considerable mineral and free gold showing. There was no question that the vein and ore shoot in the Eureka and Idaho mines were one and the same. The footwall formation was serpentine, and the hanging wall consisted of a diabase dike. The vein strikes north 77° west and has an average dip of 70° southwest in the upper levels.

The horizontal length of the ore shoot ranged close to 300 feet before the vein pinched out. This made it necessary to continue sinking the shaft and to develop additional levels to keep ahead of stoping operations. At the same time, greater hoisting capability was badly needed. This was achieved by removing the Cornish pump column from the shaft and replacing it with three Stoddard steam pumps. A new hoist with a 10" cylinder and a 16" stroke was set in place, and the pump compartment was then used for hoisting ore.

By the beginning of 1870 the Idaho vertical shaft had progressed to below the 600-foot level, by which time it was necessary to retimber much of the upper 300 feet. It was quickly learned that the dryness of the air in the mine accelerated timber decay. In addition, the serpentine in the footwall caused timber damage through the pressure exerted against the timbers by swelling ground. The number of men working on the surface and underground totaled 101, each receiving $3 per day for a 10-hour shift, with the exception of 12 men who received $2.50 per day.

New Shaft Commenced—1871

By 1871 greater hoisting capacity was badly needed to keep up with the underground production, so a new shaft was commenced 150 feet east of the vertical shaft. It measured 6½ by 20 feet inside of the timbers, and was divided into four compartments, two for hoisting ore and men, one for pumping, and one for sinking. In order to save time and to take advantage of gravity, the shaft was worked mostly from underground by both sinking and raising.

By the end of 1872 the new shaft had advanced to the 400-foot level, with only a small working raise between the 500- and 400-foot levels. A series of working raises was run, one in line with the other, from one level to the next, all the way to the surface. These raises were then enlarged to the dimensions of the shaft and timbered accordingly. Much of the shaft was timbered through stoped-out ground. The rock from the new shaft

The surface crew at the Idaho mine.

Miners and other employees at the Idaho mine

Interior of Idaho hoist room. Engine for Cornish pump.

Winding reels and flat steel cables were used for hoisting during this period.

was hoisted through the vertical shaft as ore, except the rock in the upper part of the shaft above the ore shoot.

From the 400-foot level 15" square timber was used in the shaft. Cages were to hoist tram wagons one at a time, and the same cage was used to transport men. The cages rode between two 5" by 5" wooden guides attached to the timber in each hoisting compartment.

Edwin Edmunds, shaft man, was at the bottom of the shaft when some rock caved on him. He was taken out of the mine as soon as possible and placed on a mattress in the mine office. He was treated by Dr. Simpson, but died shortly after. The accident occurred on January 4, 1872.

Surface Improvements—1873

Two hoists were set in place near the collar of the new shaft, along with a new headframe and steam boilers. One hoist engine had a 14" cylinder with a 5-foot stroke, and was used for hoisting ore. A second hoist with a 10" cylinder and a 16" stroke was used for hoisting tubs while sinking in the shaft. Flat steel cables and winding reels were used on each hoist. A large wooden building was constructed over the headframe and hoisting works.

The mill was increased from 15 to 35 stamps in order to work the mine on a larger scale. Each stamp weighed 850 pounds and dropped 65 times a minute from a height of 10 inches. To further speed the crushing, two new rock crushers were installed ahead of the stamps to reduce the rock to two-inch size. The steam boiler to run the mill used seven cords of wood every 24 hours.

The Idaho mine had risen to being first in gold production among all companies in California. During the four years of operation, 1869 through 1872, the total earnings were $1,300,823, with 41 separate dividends paid to the stockholders, which amounted to $602,950. Of the 3,100 shares of stock, 2,880 were owned in Grass Valley, whereas only 220 were owned by non-residents.

In 1873 it became apparent that a greater supply of water was needed for milling operations. Land was purchased and a reservoir was constructed at an elevation above the mine to provide 520 feet of fall. Water rights from Little Deer Creek on Canada Hill (above Nevada City) were purchased, and a ditch between the reservoir and Little Deer Creek was completed. Rights-of-way were also obtained and a ditch was extended from Little Deer Creek to Scotts Flat to bring water from Deer Creek to meet the mine's future needs. The reservoir, water rights and labor costs amounted to $12,114.

The Idaho Quartz Mining Company purchased the rights to the Schofield Mining Company's claim for $25,000. This claim was adjacent to the Idaho claim, and was bought to preclude conflict and need for legal action in the future.

Underground Operations—1873

The new shaft was opened in 1873 for hoisting both ore and waste rock, and for lowering men and supplies. The shaft continued downward, and by the end of 1873 it was advanced 35 feet below the 700-foot level. Advancement in the lower section was slower, because during the sinking operation the rock had to be shoveled into tubs. Above the 600-foot level, raises were used for shaft excavation, and the rock was caught in chutes. The old vertical shaft was sunk to the 800-foot level, where it was stopped and abandoned for hoisting purposes.

The underground operations were progressing well. The Idaho ore shoot was raking eastward in a fairly flat plane on both the 600- and 700-foot levels, therefore a greater amount of ore was able to be mined on these two levels. The Idaho mine had an excellent year, and established itself solidly as the leading quartz mine in California, and one of the most productive gold mines in the state. During 1873 there were 27,624 tons of ore milled with a gold value of $1,024,591. The dividends for the year amounted to 220 percent on the capital stock, which amounted to $682,000.

Nevada County Narrow Gauge Railroad

Businessmen of Nevada City and Grass Valley recognized the advantages of a railroad connection with Sacramento and San Francisco. A bill was submitted to the state legislature and was approved on March 25, 1874, for a railroad between Nevada City and Colfax. This would connect with the Central Pacific Railroad.

A company was formed with officers as follows: John C. Coleman, president; Thaddeus W. Sigourney, vice president; George Fletcher, secretary; J. H. Bates, chief engineer; Edward Coleman, James M. Lakenan, Niles Searls and Richard W. Tully, directors.

A contract to construct the railroad was awarded on December 4, 1874, to Moses Findley Beatty, who sublet it to Turton and Knox. The work commenced in February 1875, and was completed from Colfax to

Grass Valley in April 1876, and trains began running between the two towns. The railroad was completed to Nevada City May 20, 1876. A large procession, headed by a military band, met the incoming train in Nevada City, to the delight of a jubilant crowd of people. Judge Niles Searls delivered the congratulatory address.

The Coleman brothers' interest in having a railroad was the same as other businessmen; however, they also needed to transport large quantities of wood to operate the steam boilers at the mine. The hoists, compressors, pumps, crusher and the mill all relied on steam power. Close to 35 cords of wood were burned daily to keep the mine operating.

The Empire, North Star and other mines in the district also consumed large quantities of wood in their boilers. The result was that trees were quickly becoming scarce in this area. The Coleman brothers had bought stands of timber in the Chicago Park and Peardale areas and needed the railroad to haul wood to the mine. The Colemans were large contributors to the Nevada County Narrow Gauge Railroad.

Change in Management—1875

In 1875 Edward Coleman replaced his brother as president, and John Coleman replaced Thomas Findley as company treasurer. Findley had been suspended earlier when it was discovered that a sum of $32,490.88 was missing from the company account. By order of the trustees, a compromise was reached whereby the company agreed to accept 10 percent of the indebtedness in lieu of the full amount. Eventually $9,500 was repaid. No dividends were paid during November, and only $3 per share was paid in December because of the embezzlement.

The collar of the pump shaft and surface pump equipment were in close proximity of the large wooden hoist and mill buildings. There was concern that in the event of a fire at either building, the pumping capability could be seriously affected. To guard against this potential loss, a fireproof structure made of brick was built over the shaft collar and pump machinery.

Since the mine started, all drilling operations had been done by hand. Following the introduction of giant powder in 1869, small diameter steel drilling rods came into use, and "single jacking" produced great savings in time and money. In 1875 the company purchased five Burleigh rock drills, one of which was used in shaft sinking, and four in drifts. These drills were operated by compressed air, the compressor being located on the surface and the air conveyed underground through a six-inch pipe. The new Burleigh drills proved to be superior to hand drilling, both in time and footage of holes drilled. The company followed up with the purchase of Ingersoll rock drills, which in time phased out hand drilling altogether.

The ledge in the shaft was of good quality, with well-defined walls, having an average width of four feet of quartz until it reached 60 feet below the 800-foot level. At that point the hanging wall flattened out and the ledge became broken up into several quartz stringers. The shaft crossed out of the ore shoot, but was continued to the 1000-foot level, where the shaft stopped. The shaft measured 1,116 feet on the incline and 991 feet vertically below the collar.

The pumping capacity became borderline with the greater depth, so a decision was made to install a larger pump. The vertical shaft was the logical place for the pump, but first a considerable amount of timbering was necessary. A new Cornish pump was installed in the old shaft. The engine cylinder was 20" in diameter, with a 42" stroke.

The mine continued to be the leading gold-producing mine in California. It was in excellent condition, with drifts ahead of stopes, thus assuring close to three years of reserves. The ore shoot remained well defined on each level, with from two to five feet in width, averaging about four feet. The quartz pinched down to almost nothing on most levels after leaving the ore body. The ore shoot raked approximately 33° to the east.

The mining practice had always been to mine out each level as the mine progressed downward. All of the quartz was taken out as the stopes were carried forward, leaving few pillars or blocks of ground within the ore shoot, even where the ledge narrowed down or was a lower grade. It was deemed more profitable to work everything rather than take chances of leaving any quartz that might possibly pay.

Use of Cannon Underground—1876

A 250-pound field artillery piece was purchased and was used underground on the upper levels where the vein was quite steep. Just as in the Eureka mine, broken rock would periodically become wedged together by its own weight and form a bridge. Following the example of the Eureka operators, the cannon was used to fire an iron projectile at the obstruction and cause the rock to fall. The previous method had been to load black powder in an empty quicksilver bottle and place it under the bridged rock. The powder was ignited remotely by a long fuse. This was very dangerous, because the miners were exposed to major injury. This condition did not occur below the 500-foot level where the vein was not as steep.

A considerable amount of timbering was necessary in the main shaft during 1877. The timbers between the 200-foot level and the surface had decayed quite rapidly due to dryness in the shaft. Much of this section of timber had to be replaced. The pump shaft needed a considerable amount of timber repair because of its dryness and swelling ground.

Production increased during 1878 with ore coming from the heart of the ore body, being the 500-foot level to the 1000-foot level. The ore shoot raked away from the shaft in a flat manner. On the 1000-foot level the east drift was driven through 300 feet of barren ground before the ore shoot was encountered. Rather than sink the main shaft any deeper, it was decided to sink a new shaft from the 1000-foot level, following the ore shoot. A large station was cut out and an air hoist was installed to sink the shaft 312 feet east of the main shaft. The new shaft measured 6 feet by 14 feet, with two car compartments, a pump compartment and a ladder way (steps). The shaft followed the west edge of the ore shoot at a 40° dip. The shaft had been advanced 105 feet by the end of 1878. It may be noted that in the true sense this was a winze, although it has always been called the Canyon shaft.

Communication between the surface and underground required a workman to relay the message directly in person by riding a cage in the shaft. This took time, and in case of an injury or other emergency this extra time could be a matter of life or death. In 1878 a line was run down the main shaft and a telephone installed at each station and the Canyon shaft hoist room.

The depth of the mine had reached a point where there was insufficient air circulation in the lower levels. In order to alleviate this condition, an air shaft was started on the 500-foot level and run through to the surface. This raise would provide natural ventilation to all mine areas by the air coming down the main shaft and exhausting through the new air shaft. It would also be beneficial as a second exit in the event of fire or other emergency.

Surface Improvement—1878

By 1878 there was sufficient ore blocked out on several underground levels to be able to increase ore production and still maintain adequate ore reserves. This meant increasing the hoisting, compressed air, and

(Above) The Idaho hoisting works (10) and blacksmith shop (11) as they appeared in this 1880 illustration.

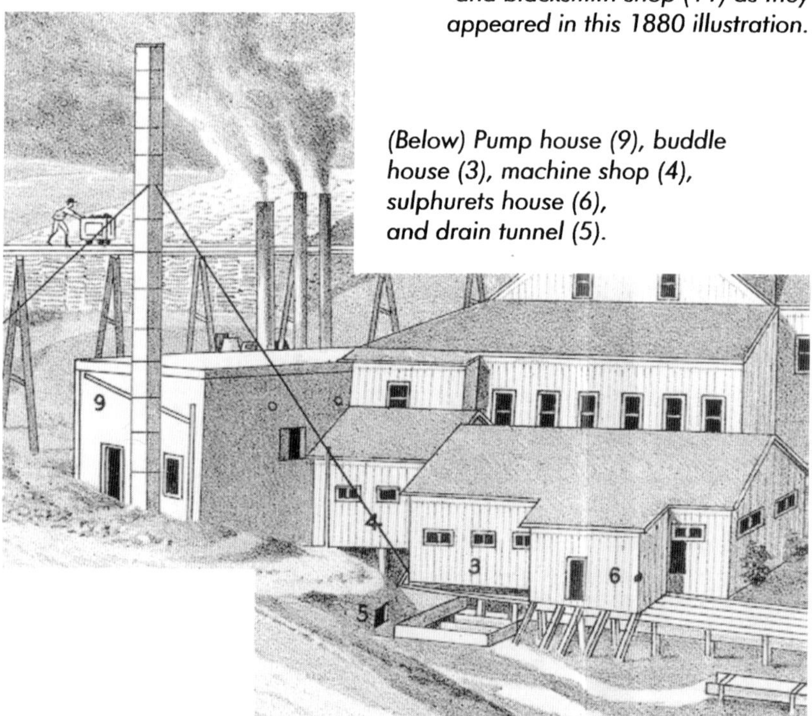

(Below) Pump house (9), buddle house (3), machine shop (4), sulphurets house (6), and drain tunnel (5).

Lithographic drawing of the Idaho quartz mine in 1880. (1) Office. (2) Mill. (3) Buddle house. (4) Machine shop. (5) Drain tunnel. (6) Sulphurets house. (7) Tailings sluices. (8) Rocker house. (9) Pump house. (10) Hoisting works. (11) Blacksmith shop. (12) Waste rock track. (13) Air shaft. (14). Water ditch. (15) Nevada County Narrow Gauge Railroad. (17) Road to Hills Flat and Grass Valley.

milling capabilities. An energetic program was undertaken to achieve these three improvements.

A National Duplex type air compressor was installed alongside the same type already in service. These were both equipped with a Myer's cutoff, and were located within the large hoist building. Each compressor was powered by an 18" diameter engine with a 42" stroke. The additional compressed air was needed to run the air-operated hoist on the 1000-foot level and several new Ingersoll drills.

When the decision was made not to sink the shaft below the 1000-foot level, the sinking compartment became available for hoisting ore. This compartment was converted by adding a safety cage for hoisting cars, thus increasing hoisting by nearly one-third. Two new boilers were added to power the additional hoist and air compressor. This made a total of four boilers, each 50 inches in diameter and 16 feet long. These were all placed in a newly constructed building covered with corrugated iron as a fire prevention measure.

The capacity of the mill was increased to 135 tons per day by the addition of another 15 stamps, bringing the total to 50 stamps, each weighing 850 pounds. The stamps were also increased from 65 drops to 72 drops per minute. To meet the power demand, an 18" diameter steam engine with a 42" stroke was added. Two second-hand boilers were placed in a newly constructed building, along with the two original boilers. These four boilers were 50 inches in diameter and 16 feet long. This building was covered by corrugated iron, just like the boiler room for the hoisting works.

The power to operate the milling, pumping, hoisting, and compressed air needs of the mine was supplied by a total of eight large steam engines and their associated boilers. Approximately 25 cords of wood were consumed daily to keep boilers operating to meet power demands.

The blacksmith shop adjoining the hoisting works was one of the busiest operations in the mine. The six forges were in constant use to keep up with sharpening picks and drill steel. The machine shop section was well equipped with such machines as a lathe, planer and drill presses to make the mine self-supporting for equipment repair.

The Idaho mill was considered among the most up-to-date in the gold mining industry. The ore was first crushed to 2" size by two Varney rock breakers, then fed into the batteries of stamps. The 50 stamps pulverized the rock to a fineness to go through a 40-mesh screen. From the mortar screens the finely ground sand and water flowed over copper aprons covered with blankets, where approximately 90 percent of the gold and sulphurets were caught in the blankets. Each blanket was changed every 20 minutes and then washed in tubs of water. The washings were passed through Atwood amalgamators, where the gold was brought in contact with quicksilver and remained there. The sulphurets passed off and were then treated in a Cornish buddle.

On February 18, 1878, a 22" x 10" timber, 22 feet long, was being lowered down the shaft when the timber came loose near the 500-foot level. George Polglase and James Buckett were working on the 800-foot-level station, and both men were struck by the falling timber. Buckett was killed instantly and Polglase died several hours later.

It is interesting to note that the funeral service for James Buckett was held at the Emmanuel Episcopal Church and the services for George Polglase took place at the Methodist Church (the churches are across the street from each other). A joint procession was formed and the bodies were taken to the Grass Valley cemetery. The men were laid to rest side by side in the same grave, with a common gravestone for the two miners.

Underground—1879

The air shaft that began in 1878 was completed in 1879. A fairly large amount of sub-surface water was struck when the surface area was being prepared where the raise was to hole through. A small pump had to be installed to divert the water from entering the mine. The air shaft proved very beneficial by increasing the downward air volume through the main shaft, and after circulating throughout the mine workings the air was exhausted up the air shaft.

As the underground workings increased in depth it was found that the boiler capacity for the Cornish pump engine was insufficient. A boiler could be spared from the hoisting works, so it was moved to the pumping works. The boilers for the pump unit were placed in a newly constructed boiler room that was also covered with corrugated iron.

During 1879 the Canyon shaft had progressed downward 175 feet to where the 1100-foot level was established. The east drift was driven 188 feet, and the quartz was low grade, just as had been anticipated, except for the last 30 feet, which was of fair quality. The blasting smoke did not clear away from this level very well, and a fan driven by compressed air was used until a raise could be put up to the 1000-foot level. Natural air circulation was then provided and the fan moved to the east drift on the

1000-foot level. A crosscut was run in the footwall on the 1000-foot level to explore a quartz stringer, but nothing of value was found.

All of the ore had been mined from the 600-foot level and above by the end of 1879. During the year 32,370 tons of quartz was milled, and of this 1,225 tons came from the 700-foot level, 8,937 tons from the 800-foot level, 6,718 tons from the 900, and 10,302 tons from the 1000-foot level. This quartz yielded 27,160 ounces of bullion with a value of $499,380. The entire product of the mine since the year 1869 was $5,699,742, and the dividends, 124 in number, had totaled $2,703,200.

The Idaho mine remained the richest and most famous mine in the state. Its surface plant and equipment ranked among the best in the mining industry. No other gold mine had paid dividends so regularly and so well.

Underground—1880

The above-normal rainfall during April 1880 caused considerable flooding for the first time in the mine's history. Surface runoff water, entering the old workings of the Eureka mine, flowed into the upper workings of the Idaho mine and continued on down and finally ended up in the bottom of the Canyon shaft. The water filled up in this shaft and flooded the 1200- and 1100-foot levels. Previously, water from the Eureka mine was always caught by the pumps on the 700- and 800-foot levels of the pump shaft. In this instance, however, the amount of water was too great for the pumps to handle. The mill and underground operations were suspended until the water was pumped out.

As the mine attained greater depth, so were costs increased proportionately. First the pumps were found to be inadequate to handle the winter rains and still keep up with the daily mine water accumulation. Larger pumps were installed on the 700-foot level to take care of the Eureka mine water as an interim measure. The pump shaft needed timber repair in several areas, so it was deemed best to install a steam capstan to assist in this work. This consisted of two 8" steam engines, with drum and all appendages necessary for its safe and efficient working.

The air circulation in the Canyon shaft at the 1100- and 1200-foot levels became inadequate. A large exhaust fan was then installed in the air shaft near the surface to increase the volume and velocity of fresh air. The exhaust fan was operated by a small Pelton wheel, fed by high-pressure water taken from the pipeline supplying water to the mill. The pump and exhaust fan located at the top of the air shaft were the first instances in which the company used water in place of steam for motive power.

There were serious concerns during 1879 and 1880 that the profitable days of the mine were drawing to a close when the 1000-foot level had run into comparatively barren ground. The vein had narrowed and the values were erratic. Then the vein on the 1100- and 1200-foot levels widened and the stopes began producing good ore again.

The mining costs increased when it became necessary to sink the Canyon shaft, following the ore shoot in its downward dip and eastern rake. This meant double handling of ore from below the 1000-foot level, and then tramming to the main shaft, where the ore was hoisted to the surface. The development work was being pushed as fast as circumstances would permit. A second air hoist with a 10" cylinder and a 24" stroke was installed on the 1000-foot level Canyon shaft station, to be used exclusively for sinking in the shaft.

By the end of 1880 the 900- and 1000-foot levels were worked out, and the 800-foot level had only a few months of ore left. Most of the ore now was coming from the 1100- and 1200-foot levels. The overall tonnage dropped from 32,370 tons in 1879 to 11,611 tons in 1880, and gold production was more than cut in half, from $499,379 in 1879 to only $226,078 in 1880.

New Cornish Pump

By 1881 the Idaho mine had reached a vertical depth never achieved by any other gold mine in the state. Stopes above the 1000-foot level had been worked out, and the future of the mine relied directly upon the ore from the levels in the Canyon shaft. Sinking operations in the shaft had been interrupted on several occasions by flooding from winter storms. Steps were then taken to provide an auxiliary pumping system.

Management contracted with Prescott, Scott and Co. of San Francisco to supply a large Cornish pump, with its engine having a 20" cylinder with a 48" stroke. This was a high-pressure engine with balance pocket values and controlled by an adjustable cutoff. The fly wheel was cast in a local foundry and was 18 feet in diameter, weighing 28,000 pounds. The spur wheel was 14 feet in diameter, having a 20" face with v-shaped teeth. The spur wheel was made in halves in order to pass through the Greenhorn tunnel while being transported to the mine on the narrow gauge railroad.

Among the more impressive things about the early days of under-

FRONT VIEW OF CORNISH PUMP SIDE VIEW OF CORNISH PUMP

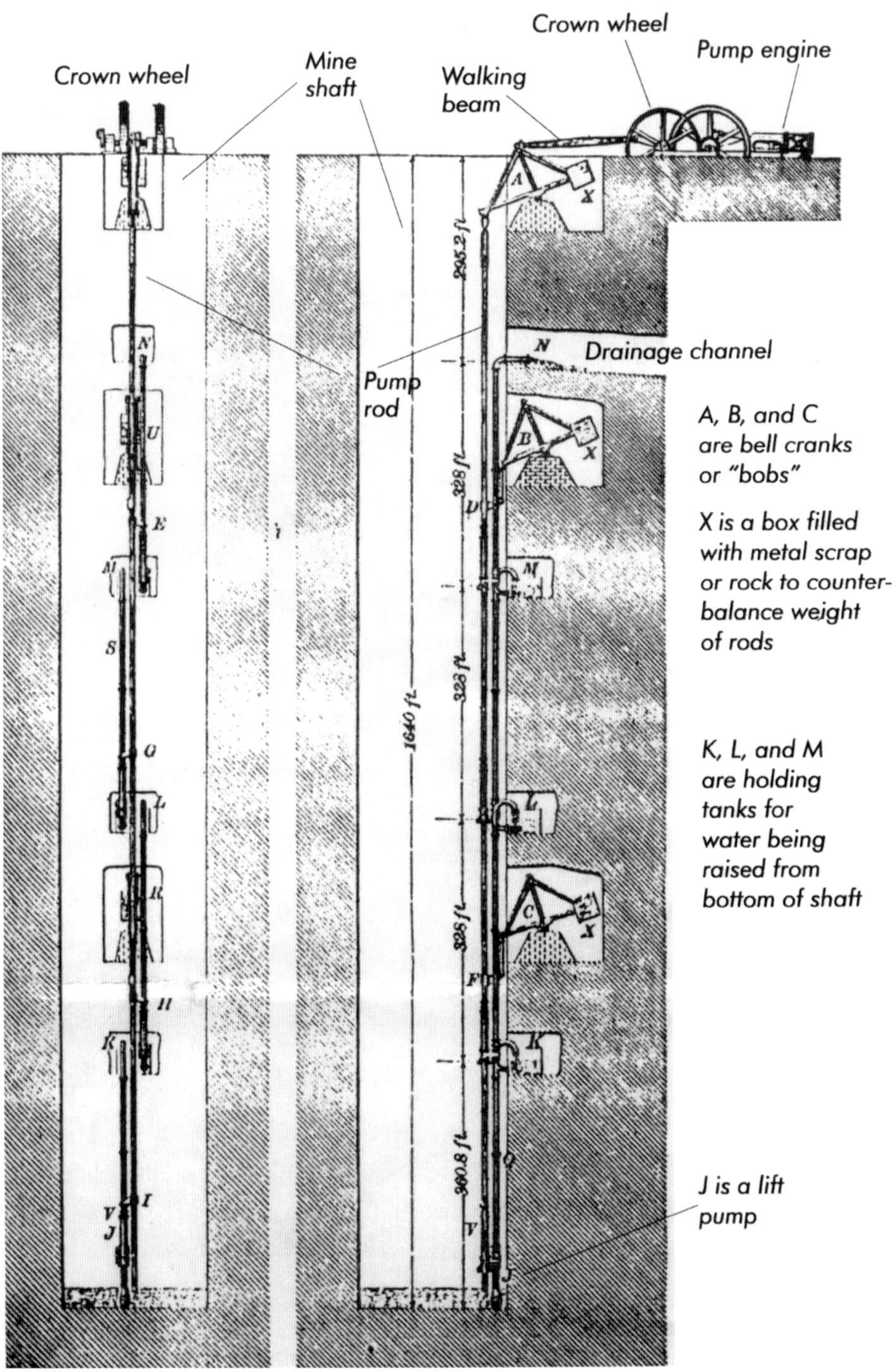

ground gold mining were the Corish pumps and their heavy pump rods. As cumbersome as they were, these pumps allowed the mines to reach depths that would not have occurred otherwise.

The preparation for the foundation was a large undertaking in itself. Master mechanic J. V. Pitchford had drawn up the specifications for the pump, bob, and foundations, and was also fully in charge of the installation. The ground was excavated to a depth of 28 feet down to solid rock, and on this a solid bed of granite was laid in cement for the engine foundation and bob pit. The foundation was 63 feet in length, 28 feet in depth, and in width one portion was 16 feet and the remainder 33 feet. The capstones for the engine were dressed granite from the quarries at Penryn. The entire excavation for the stone work was 38,000 cubic feet. The steel used to strengthen the foundation was 12,000 pounds, and the anchor bolts to secure the equipment were 2½ inches in diameter. The entire pump unit was housed in a building 47 feet wide, 44 feet long, and 30 feet high.

The pump rod was in three sections, extending from the surface to the sump below the 1200-foot level. The upper section of the pump went down the Idaho shaft to the 1000-foot level, a slope distance of 1116 feet. The shaft dipped 73° for 175 feet, then averaged 55° from that point to the 1000- foot level. The rod consisted of 69 timbers, each 14" x 14" and 16 feet long. The timbers were spliced together, end to end, by heavy cast-iron plates bolted on opposite sides of the timbers. The pump rod was attached to the bell crank of the large Cornish pump. The bell crank was always called the "bob" by the Cornish miners. The rod moved up and down on rollers in the shaft. The bell crank was 36 feet long and 16 feet high. The pins for connecting the parts were 6½ inches in diameter, and the bob turned on a 10" diameter wrought-iron shaft. The bob was the largest of any in the state, and the entire pumping unit was considered to be the best in California at that time.

The second (middle) section of rod ran horizontally for 310 feet along the 1000-foot level, from the Idaho shaft to the collar of the Canyon shaft. This was done by a combination of wooden rods, bell crank levers, "bobs" and chains. Edward Coleman sent to England for a man to come to the mine and engineer the pump rod movement between the two shafts.

The third (lower) section of the pump rod went down the 40° dip of the Canyon shaft and was extended as the shaft was being sunk. This lower rod moved up and down on rollers, as in the Idaho shaft.

This authentic Cornish pump can be operated by visitors at the North Star Power House mining museum in Grass Valley.

Water was pumped up the Canyon shaft by bobs attached to the pump rod. The water was then pumped from the Canyon shaft to the large sump on the 800-foot level near the vertical shaft. The storage capacity of the 800-foot-level sump was greatly enlarged to hold water from the lower levels, and also contain winter storm water from the Eureka mine. The new sump measured 16 feet wide by 14 feet deep by 60 feet long.

A separate 12-inch pump located at the collar of the pump shaft raised the water from the 800-foot level to the 700-foot level. From there the 14-inch pump raised the water up the pump shaft to the drain tunnel. There were clack valves in the pipe columns to prevent the water from draining back down the pipe.

The Cornish pump was certainly one of the most important pieces of machinery used during the early years of underground mining.

CHAPTER 3: IDAHO MINE – PART II

Following the developmental year of 1880, 1881 looked considerably better. It showed that the mine could still yield profitably for years to come. The vein on the 1100-foot level, as it extended eastward, became well defined and carried good values. The vein in stopes above that level was also strong, even under the barren ground that appeared on the 1000-foot level. The vein on the 1200-foot level appeared strong and uniform in size, with the stopes above providing good ore.

Of the 27,600 tons of ore hoisted during 1881, 9,754 tons came from the 1100-foot level drift and stopes, 15,528 tons from the 1200-foot level drift and stopes, 1,978 tons from the 800-foot level, and 340 tons from shaft sinking. The 1300-foot level was advanced 124 feet by the end of the year, and the shaft sinking continued toward opening the 1400-foot level. The shaft was beginning to crowd the hanging wall, and it became necessary to turn the incline slightly to the left in order to continue following the footwall.

The year 1881 was one of great prosperity, being the best since 1874, and the third best since the mine commenced paying dividends. In 1873 the gold production was $1,010,612; in 1874 it was $669,023; and in 1881 it amounted to $640,108. This was $199,762 greater than the yield of 1880, showing an improvement of the mine in its deeper workings.

Boiler Explosion—1882

Back in 1879 a small Pelton wheel had been put in use to operate a small pump near the collar of the air shaft. Water to run the Pelton wheel was taken from the pipeline that supplied water to the mill. Mine management could see the advantages of water power, and considered changing to it entirely. Steam had been the source of power by all of the local mines since their beginnings, and the consumption of wood to feed the boilers was having a deleterious effect upon the forest. The Idaho mine alone was consuming 30 cords of wood daily to keep the boilers running.

In the year 1882 the California state legislature prohibited further hydraulic mining operations, thus allowing water storage dams and canals to become available for other uses. As the Pelton wheel had recently been perfected, it was an opportune time for the Idaho company to look toward changing motive power from steam to water. The company already had a large reservoir, and a ditch system from the reservoir to Deer Creek, near Scotts Flat. In addition the company owned water rights on Little Deer Creek.

The Idaho Quartz Mining Co. entered into a contract with the South Yuba Water Co. to supply water for the mine's needs. Orders were placed for a number of sizes of Pelton wheels necessary for the changeover. A 22" diameter wrought-iron pipeline was started between the mine and the reservoir, which would deliver 360 miner's inches of water to the mine under 542 pounds of pressure.

On August 4, 1882, a boiler explosion occurred at the hoisting works, tearing out the end of the building, the boiler landing 40 feet away. Luckily, no one was injured by the incident. The immediate result of the explosion was to interrupt the hoisting of ore and operation of the new large pump that had recently been put in service. The miners who were working underground on the night shift had a long, strenuous climb up ladders to reach the surface. The steam connections were quickly repaired, and the hoists and pump were back in service with a loss of only two underground shifts.

At about 8:00 a.m. on September 26, 1882, a fatal accident occurred in the mine. Thomas Williams, Frank Johns, Richard Johns and Richard Carter were at work at the bottom of the Canyon shaft when a large cave of rock came thundering down the shaft from near the 1100-foot level. The rock dislodged some timbers in its path, and the four men were caught by the mass. Thomas Williams was killed instantly when he was struck by a timber. The other three men were covered by rock and debris. Richard Eva and E. Magor were the first on the scene to free the injured. Richard Carter sustained two broken ribs and a fractured jaw, whereas

Night shift at the Idaho mine.

Day shift at the Idaho mine.

The Pelton Water Wheel. In 1895 the manufacturer boasted that it had sold 4,000 wheels, with power "aggregating more than 300,000 horsepower."

National duplex air compressor and hoists, all run by water power. Water was supplied by the South Yuba Water Company.

the other two received only multiple bruises. It was fortunate that their injuries were not more severe.

At the end of 1882 the 1100-foot level drift had been driven 1044 feet from the incline, but the last 100 feet had crossed through the ore shoot and was in barren ground. All the stopes on this level had been worked out. The 1200-foot level was in 861 feet from the incline, and the vein had been stoped through to the 1100-foot level for 700 feet from the incline. The 1300-foot level drift had been driven 422 feet from the incline, with the ledge averaging 2½ feet in width in both the drift and stopes. The grade of ore was about average quality. The Canyon shaft was down 221 feet below the 1300-foot level and a station had been cut for the 1400-foot level.

During the year, 27,539 tons of ore were hoisted, of which 5,142 tons came from the 1100-foot level, 13,680 tons from the 1200-foot level, 8,525 tons from the 1300-foot level, and 191 tons from the Canyon shaft. This yielded 31,605 ounces of bullion and concentrates, for a total value of $565,386. The ore averaged $20.64 per ton.

Changeover to Water Power—1883

During most of 1882 and the spring of 1883 the company expended a large amount of money to make the changeover from steam to water power. Three 8-foot-diameter Pelton wheels, four 6-foot Pelton wheels, and one 9-inch Pelton wheel were installed to replace their respective steam counterparts. The boilers were left in place as a backup in the event of a water shortage or other water-related interruptions.

The 22" diameter wrought-iron pipeline, 8,700 feet in length, was completed between the mine and the company reservoir. Water was purchased from the South Yuba Water Co. and delivered from Scotts Flat through the mine's ditch system. Three hundred and fifty miner's inches of water were supplied, 30 inches of which were used to meet the needs of the rest of the mine. The use of water for motive power was a new departure in local mining operations, the Idaho company being the first to make this change.

On July 8, 1883, an 8-foot Pelton wheel, used to drive the pumps, burst into pieces. The fragments caused minor damage to the hoist building roof, but fortunately no injuries occurred. The wheel was turning at 285 rpm, driven by 60" of water at 542 feet of pressure. The theory for the failure was that the rim thickness was too light for the pressure and velocity under which it was running. The standby steam boiler was activated, and within hours steam power continued until a stronger Pelton wheel was manufactured.

The ore storage bin in the mill was enlarged and self-feeders for the batteries were added. The tracks bringing ore to the mill were not high enough to give adequate room for a bin of proper dimensions. To gain the additional storage space, it was necessary to move the rock breakers to a new location and then hoist the crushed rock into the bin.

The quality of ore milled during 1883 was much below that of the previous year. By the end of the year the 1200-foot level had reached 1,106 feet from the incline, the last 200 feet of which was in very good ore. The 1300-foot level drift was in 713 feet with average quality ore. The 1400-foot level was in 389 feet with a good-sized vein, and the Canyon shaft was down 168 feet below the 1400-foot level.

During the year 23,575 tons of ore were crushed, of which 8,140 tons came from the 1200-foot level, 15,513 tons from the 1300-foot level, 4,587 tons from the 1400-foot level, and 332 tons from the shaft. The average yield was $12.76 per ton, and the total gold produced for the year was valued at $364,600. The dividends for the year amounted to only $34,500, lowest since 1869. This was attributed mainly to a lower grade of ore mined, but another factor was the initial expense for the changeover to water power. For the first time in the mine's history, there were five consecutive months in which no dividends were paid.

Year 1884

Entering the year 1884 the development work remained well ahead of stoping operations. Sinking continued in the Canyon shaft, and the 1500-foot level was established. Air doors were installed on all levels off the Canyon shaft to provide for good air circulation throughout the lower workings. The air circulated down the Canyon shaft to the bottom level, and then flowed up raises, from one level to another, until it reached the 1000-foot level.

During the year the ore from the drift and stopes on the 1200-foot level had been exhausted. The vein in this level gave out at 1,185 feet from the incline. Drifting and stoping operations continued on the 1300-, 1400- and 1500-foot levels.

The shaft sinking averaged 28 feet per month, and by the end of the year it was down 119 feet below the 1500-foot level. There were 31,143 tons of ore crushed during the year, of which 1,637 tons came from the

1200-foot level drift and stopes, 11,268 tons from the 1300-foot level, 18,098 tons from the 1400-foot level and 143 tons from the incline.

The year 1884 was considerably more profitable than the previous year. The grade of ore was $7.28 per ton higher, and more than 3,000 additional tons of ore were milled than in 1883. Gold production amounted to $561,895, and dividends were $271,250. The average value of the ore for the year was $18.04 per ton, as against $12.76 per ton in 1883, and the cost of milling and mining was $7.86 per ton, while in 1883 it was $8.71 per ton.

The conversion from steam to water power proved beneficial, from both an operational and economic standpoint. It was estimated that the company would realize a savings of over $30,000 each year. The number of employees was 220. The muckers and mill workers received $2.50 per day, and miners, blacksmiths, hoistmen and mechanics $3.00 per day.

Year 1885

The mine used 40 miner's inches of water (580 gallons per minute) for fire protection, milling operations, and other surface requirements. This was separate from the high-pressure water used to operate the Pelton wheels. Low-pressure water needs were met during the winter months by purchasing low-cost water from near the headwaters of Wolf Creek. In the early summer months the water came from the company's water rights on Little Deer Creek. In late summer and early fall months the water came from Deer Creek, and was purchased from the South Yuba Water Co.

Operations progressed fairly well during 1885. The mill was kept busy, with 35 stamps running constantly. The 1300-, 1400 and 1500-foot levels produced fair milling ore, and the drift on the 1600-foot level was being driven, but had not yet reached the ore shoot. The shaft sinking continued and was close to the 1700-foot level elevation.

The practice of mining and milling all quartz within an ore shoot, even when the vein narrowed down, often resulted in a low overall grade of rock. To maintain the normal drift size, or a workable width for working purposes in a stope, the ore automatically became diluted with lower grade wall rock. This was not altogether bad, because this mining was still profitable and prolonged the life of the mine.

The mine produced 30,518 tons of ore during 1885, which was only 634 tons less than the previous year. Because of the lower grade of rock, the production yield was $370,197, down $191,000 from the production of 1884. Dividends were commensurately lower. Although the grade of ore and dividends might drop off, the mine had always bounced back and responded with dividends to the satisfaction of its stockholders.

Year 1886

In June 1886 the Idaho Quartz Mining Co. declared its 200th monthly dividend. In over 17½ years, only seven monthly dividends had been missed. A record of this kind is deserving of more than passing notice, as no other gold mine on the North American continent had ever paid this number of dividends. Since 1869 the mine had produced over eight million dollars, with dividends exceeding $3,750,000. The rich ore shoot was the main reason for such lucrative monthly dividends, but intelligent management also played an important role. The Coleman brothers were always progressive, maintaining the mine and equipment in first-class condition, while at the same time modernizing operations as required.

The ore shoot continued in its easterly rake at about 40°, and it appeared very likely that it would ultimately cross over into the Maryland ground. With each successive level driven, the ore shoot veered to the east, away from the Canyon shaft. It was decided to continue sinking the shaft in hopes of discovering another ore shoot, or a new vein system.

The year was very profitable, with little need for major repairs, or for new equipment. There were 29,244 tons of ore milled with a value of $547,570. Dividends to the stockholders amounted to $263,500.

The Idaho milling process was designed to extract as much free gold and concentrates from the ore as possible. Concentrates were subjected to four separate concentrations; even so, a small percentage of fine gold still encased in the sulphurets escaped in the tailings.

Shortly after the milling operations had begun in the early 1870s, the company agreed to allow two separate groups to rework the mill tailings on a share basis. Over the years the work performed was beneficial to both the company and the men doing the work. Neither process recovered a lot of gold, but it was additional profit that otherwise would be lost.

Year 1887

In 1887 the mine had reached a depth of 2,790 feet on the incline, or a vertical depth of 1,790 feet from the surface. It became necessary to place additional bell cranks ("bobs") underground to reduce the space between bobs, which in turn enhanced the efficiency of the Cornish pump system. It was also necessary to replace some of the pump rod sections where undue strain had caused some damage to the rod. The Cornish pump removed an average of 21,600 cubic feet of water per day.

The timber crews were kept busy replacing timber in the drifts and shafts. Replacing timber in the main shaft and the pump shaft was an ongoing job, especially in the upper sections of these two shafts. The timber was exposed to the continuous footwall swelling. The dry condition in the upper part of the shafts caused the timber to decay quicker than normal.

The year proved profitable for the company. There were 26,686 tons of ore taken from the 1400-, 1500- and 1600-foot levels. The mill produced $492,638 from this ore, of which $235,600 was paid in dividends.

Year 1888

Several major changes were made in the mill during 1888 to upgrade its efficiency. The old system of amalgamation, with blankets and blanket washings, was abolished, and in its place the latest improved system of silver-plated aprons was installed. The old Cornish buddles were also removed and replaced by ten Frue concentrators and six Triumph concentrators. The latter were equipped with the improved Blasdel belt. The mill was considered to be in first-class condition in every respect.

On April 21, 1888, John C. Coleman, treasurer of the Idaho mine, was seriously injured at the corner of Bennett and East Main streets in Grass Valley. While on his way to work at the Idaho mine, the king bolt of the buggy in which he was riding broke and detached the front axle, front wheels and horse from the vehicle. Coleman was thrown to the ground nearly head first. Sheriff George Lord was nearby and he took the injured man to his home, where Dr. William C. Jones treated him for a dislocated left shoulder and numerous cuts and bruises.

The mine continued to produce well during 1888, with 26,644 tons of ore milled. The ore came from the Canyon shaft and the 1300-, 1400-, 1500- and 1600-foot levels. The production yield was $603,695, which was the fourth highest in the mine's 21-year history under the Coleman brothers. The dividends, amounting to $325,500, were the second highest during the same period. The Idaho mine remained the leading underground gold producer in the state.

There were 173 miners and car men working underground, 17 men

The Idaho stamp mill with amalgamating plates in the foreground.

Concentrtators in the Idaho mill.

working in the mill, and 16 working on the surface. The wages paid were $3 per day for underground and mill men. The wages for blacksmiths, hoistmen and bosses ranged from $3 to $5 per shift.

The old board of directors was reelected, consisting of John C. Coleman, Edward Coleman, Miles P. O'Connor, Eugene C. Creller, and George C. Hill. Officers elected were Edward Coleman, president and superintendent, Eugene Creller, vice president, George Hill, secretary, and John Coleman, treasurer.

Mine Catastrophe—1889

At approximately 5:30 p.m. on the afternoon of June 4, 1889, a tragic event happened on the 1000-foot level. Steve Uren and James Bastian discovered a fire in the timber above the collar of the Canyon shaft. Miners were changing shifts, and fortunately all of the day-shift men had been hoisted from the lower levels.

Superintendent Edward Coleman and foreman John Carter had just arrived home when they received word of the fire. Both returned to the mine immediately. They went down to the 1000-foot level and found several miners attempting to put out the fire. The effort to extinguish the blaze was continued without success.

When the men began feeling weak from the presence of carbon monoxide and insufficient oxygen, they left the area, but by the time they had reached the main shaft their candles had gone out from lack of oxygen in the atmosphere.

Many miners were in a weakened condition, and they were helped onto the cage by those still strong enough to assist them. Francis H. Carter Jr. was fatally injured while the cage was enroute to the surface; he fell to the side and came in contact with the shaft timbers and was killed instantly. Several of the miners were unconscious when the cage reached the surface, including superintendent Coleman.

John C. Ralph and Thomas Dunstan were not on the last cage, and an immediate attempt to get back to them failed. By this time the smoke had begun rising up the shaft. An airtight bulkhead was then constructed across the shaft below the 900-foot level to keep the mine clear of gas and smoke above that point.

Two submarine divers were brought in to retrieve the two bodies from the contaminated atmosphere of the mine. Two attempts failed and the plan was abandoned. Heat generated by the fire was too great for men in heavy rubber diving suits to endure. In addition, the suits were designed for underwater use and did not respond satisfactorily under atmospheric pressure conditions in the mine.

As soon as it was demonstrated that reaching the 1000-foot level was impossible, Coleman decided to fill the mine with carbonic acid gas to smother the fire and neutralize the poisonous gasses. A large brick furnace was constructed near the shaft collar and a fan erected to force the carbonic acid gas down the 8" pipe through the shaft bulkhead. It was estimated that it would take 1,000 bushels of charcoal each day, and about three weeks to clear the mine.

On July 7 exhaust fans were placed to draw air up the mine shaft. The bulkhead was removed and continuous air testing was performed by lowering a candle light down the shaft on an empty cage. Men were lowered to the 1000-foot level station when the air was determined to be safe for breathing. The bodies of John Ralph and Thomas Dunstan were then brought to the surface.

A fire underground is always very serious, because the timber is damp, causing a fire to smolder and give off considerable amounts of smoke and carbon monoxide. Oxygen deficiency occurs very quickly, due to space constraints within underground workings. If the miners and supervision in this disaster had realized the seriousness of a mine fire, the three deaths could have been prevented. There had been ample time for everyone to evacuate the mine.

The fire made underground workings inaccessible for 43 days. After it was safe to reenter the mine, it was determined that the Cornish pump system had received no damage, and pumping began immediately. It took ten days to remove the accumulated water from the Canyon shaft workings. In the meantime, the damaged timbers at the 1000-foot level Canyon shaft station were replaced. The two air hoists luckily were not damaged, although the wooden bed plates beneath one hoist had burned and needed to be replaced. Mining resumed in the lower levels during the last week in July. Despite the loss of production time, the mine ended the year on a profitable note. There were 21,448 tons of ore milled, yielding $407,385, with $178,250 paid in dividends.

Year 1890

The first water shortage since the company changed from steam to water power occurred in January 1890. The ditches of the South Yuba Water Co. became clogged with snow and ice, which greatly diminished the water flow. There was not enough water to supply both the Idaho and

Empire mines, so the water company opted to deliver the bulk of the water to the Empire mine. Their rationale was that the Idaho mine had firewood with which to keep its reserve steam-powered machinery going, while the Empire had none. The water company was able to clear its ditches, and within a week the water flow was back to normal.

On March 8 the Idaho Quartz Mining Co. brought suit against the Maryland Gold Quartz Mining Co. to determine the boundary line between the two mines. The question of the boundary had never been in dispute before, since the line was supposed to have been settled 20 years earlier. The line had been established then in conformity to what was considered to be the lode line.

The recent underground surveys, however, demonstrated that the true course of the lode was on a line of surface croppings some 400 feet to the north. The outcropping had always been regarded as another vein upon which the Maryland company also had a location, paralleling a portion of the Idaho location. The Maryland company had previously done work upon the outcropping on the presumption that it was a parallel vein. In regard to it, there had been no previous contention between the two companies.

In the suit brought by the Idaho company, it was assumed that the vein in the cropping was not a parallel vein, but the true course of what had been known as the Eureka-Idaho lode. If this were to be established, and the boundary line fixed in accordance with the complaint of the Idaho company, then a line drawn perpendicularly from the surface down to the lower workings of the Idaho mine would, at a depth of 1,500 feet, cut into the Maryland ground about 200 feet. At that depth the vein was strong and the ore quite rich.

In May 1890 the vein formation widened out to 12 feet when the Canyon shaft reached the 1700-foot level. The quartz showed free gold and there were wide bands of sulphurets that also showed free gold. This was first thought to be the apex of another ore shoot, but the vein quickly narrowed as drifts were started on both sides of the shaft. Considerable good ore was taken where the shaft penetrated this area, along with stoping close to the shaft.

Mining continued in the Canyon shaft and in the 1400-, 1500- and 1600-foot levels, but the drifts were stopped when the Idaho end line was reached. The Eureka-Idaho vein was rich in gold values on all three levels where it entered the adjoining Maryland ground.

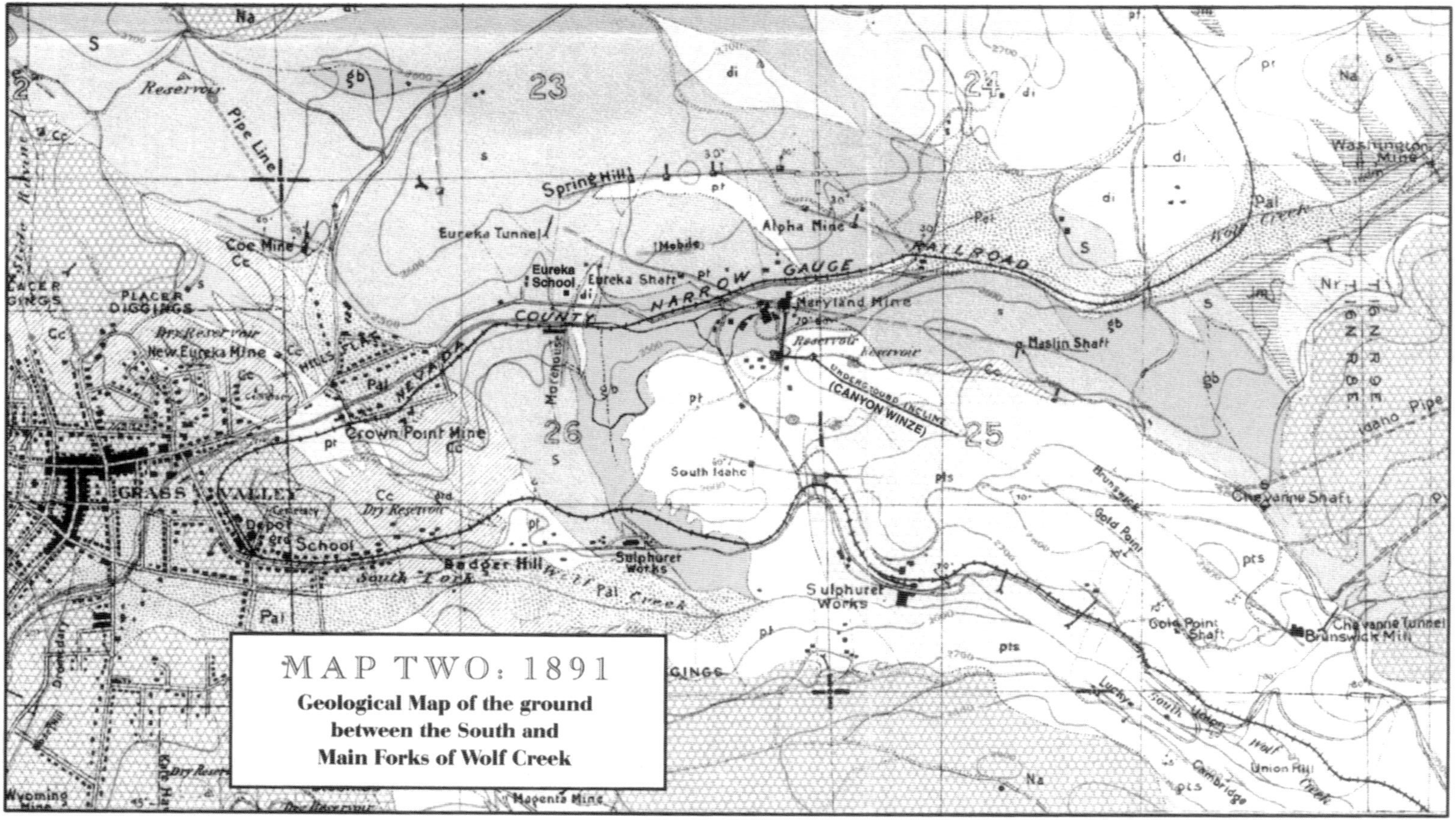

MAP TWO: 1891

Geological Map of the ground between the South and Main Forks of Wolf Creek

The case had proceeded to trial, and propositions for a compromise had been discussed but could not be agreed upon. In the rulings of the court, both sides would have gained apparent advantages, but it was evident that if the suit continued, whatever the decision of the lower court, the case would have been carried on to the California Supreme Court. Years would have been required for a final decision to be reached. In November the Idaho and Maryland companies arrived at a compromise regarding the boundary line, this ending the litigation. Counsel for both sides and individual major stockholders of each company had recommended a friendly compromise.

The terms agreed upon were that the Idaho company should have 3,100 feet on its lode line from the Eureka boundary, and at the end of that distance a true north and south line should be drawn. This gave the Idaho company about 100 feet of the ground in dispute. In this adjustment the Maryland company retained the rights to its tunnel, which was located within the ground conceded to the Idaho company. The hoisting works and improvements of Maryland company remained theirs, as they were just east of the compromise line. In addition, the Idaho company agreed to pay $6,500 and legal expenses to the Maryland company.

The conceded ground enabled the Idaho company to extend the drifts on the 1400-, 1500- and 1600-foot levels eastward an additional 100 feet in relatively rich ore. Drifting on these three levels had been stopped awaiting the outcome of the suit.

The Maslin shaft was conceded to the Idaho company as part of the compromise agreement. This prospect shaft, which was about 40 feet deep and showed a strong ledge, had been sunk by the Maryland company many years earlier. After determining that the Maslin shaft was on the same lode as the Eureka-Idaho vein, the Idaho company decided to sink the Maslin shaft deeper to explore the vein.

The grade of ore mined during 1890 was below that of the previous year, as is reflected in the production figures. There were 20,321 tons of ore milled, with a yield of $268,904. Dividends amounted to $52,700.

Year 1891

The Maslin shaft sinking operations continued during 1891, and reached a depth of 160 feet before being stopped. There were hopes that a chimney of the Eureka-Idaho-Maryland ore shoot would extend upward far enough to develop new ore. The vein, however, was narrow and not conducive for making ore.

Considerable dead work was done during the year, which included sinking and drifting. The Canyon shaft was advanced downward to a depth of 2,080 feet vertically below the surface. This measured 3,000 on the incline along the dip of the vein. The 1800-foot level was driven 195 feet east of the shaft, but the vein was too narrow for profitable mining. The 1900-foot level was driven 460 feet to the east of the shaft, and again the vein was too narrow to mine. The amount of waste rock generated from development work in the lower mine workings cut into the time for hoisting ore. This in turn reduced production during the year.

The compromise agreement in 1890 allowed the Idaho company to begin drifting and stoping for an additional 100 feet on three levels. This was very important to the company, as it added some time to the life of the mine. There was no other ore in sight than from these three levels.

Good ore continued to be taken from the 1400-, 1500- and 1600-foot levels. The tonnage milled during the year was only 16,759 tons, but the yield figure of $314,037 reflected a good grade of ore mined. The dividend was only $10,250, the lowest since the mine began in 1869. Reflected in the dividends was the difference in the cost of development work in the lower part of the mine below the 1600-foot level. Above the 1600-foot level the development costs were offset due to the development rock being hoisted as ore, whereas below the ore shoot the rock was hoisted as waste and there was no return to offset the development costs.

Year 1892

The company accelerated the development work in 1892, because finding a new source of mill rock became critical. The lower mine workings of the Canyon shaft area below the 1600-foot level had been disappointing. The available known ore in the mine was quickly being depleted.

The east drift on the 2000-foot level was driven 195 feet from the Canyon shaft, and the west drift was driven 350 feet from the Canyon shaft. The vein was very weak and the assay values low. The vein in the lower workings contained a greater amount of sulphurets than was found within the ore shoot above.

Much of the ore mined during the year came from the area between the original Idaho boundary line and the compromise line agreed upon in 1890, involving the 1400-, 1500- and 1600-foot levels. In addition, ore came from remnants of stopes on these three levels. All ore possibilities had been exhausted by the end of the year. The ore milled during the year was 16,500 tons, with a return of $248,270. Dividends amounted to $57,470.

The Idaho mine was ranked the leading quartz mine in California from 1869 through 1892. There were 567,029 tons of ore mined, yielding a total of $11,470,573 during that period. The average yield per ton was $20.23, and the total dividends paid were $5,008,433.

The combined weight of the Cornish pump rod, from the surface to the sump below the 2000-foot level, was in excess of 345,000 pounds. This rod was partially counterbalanced by the water in the pump column and the huge bob on the surface.

Imagine a mass of wood and iron working up and down in unison, while strung out over half a mile through two shafts that were 310 feet apart. The combined weight of the pump rod, from top to bottom, was over 345,000 pounds. The rod moved up and down six feet with each stroke, six times a minute, day and night.

Idaho Maryland head frame, hoist equipment, guides and cages. Air compressor is in the foreground.

Three-shaft compartments with 5" by 5" guides. Cages for hoisting ore cars can be seen at shaft collar.

Samuel Packwood Dorsey.

Victor L. Dorsey, son of Samuel.

Dr. William C. Jones

Dr. George O. Willis

CHAPTER 4: MARYLAND MINE

The Maryland mining claim was located in June 1865 by Samuel Packwood Dorsey, the local Wells, Fargo Express agent. The claim measured 2,000 feet by 600 feet, and its western boundary adjoined the eastern boundary of the Idaho claim. Although there was no visible evidence of any quartz outcropping along this claim, Samuel Dorsey indicated that he could follow the course of the Eureka-Idaho vein by the plant life on the surface. He adopted the name Maryland from his family's longtime association with the state of Maryland.

No work was done until 1880, when the Maryland Quartz Mining Co. was formed, with incorporation following on May 13, 1880. The directors named were Samuel P. Dorsey, John C. Coleman, Edward Coleman, Samuel Bethel and Stephen Moore. The elected officers were Samuel Dorsey, president; John C. Coleman, treasurer; and Samuel Bethel, secretary. The amount of capital stock was ten million dollars, divided into 100,000 shares with a par value of $100 each.

The early history of the Maryland mine is limited. In 1880 and 1881 a tunnel was driven 675 feet to the vein, and then drifts were run short distances on the vein, both east and west. A hoisting works was erected about 3,000 feet east of the Idaho shaft, and a shaft sunk to a depth of 350 feet. The Maryland shaft intersected the drain tunnel at 150 feet below the surface. The shaft was located close to the western boundary of the Maryland claim, and a very short distance from the east end line of the Idaho claim. The collar of the shaft was at an elevation 220 feet higher than the collar of the Idaho shaft. The extension of this shaft would intersect the Eureka-Idaho ore shoot at a very advantageous location. During the long years of inactivity by the Maryland Gold Quartz Mining Co., the neighboring Idaho mine was producing millions from the Eureka-Idaho ore shoot. Samuel Dorsey repeatedly refused the most tempting offers for his property. He preferred to hold and prospect the mine, believing that in time he could find a vein as good as that of the Idaho.

As the Idaho mine workings came close to the Maryland claim, the Maryland company found itself in a unique position, knowing that the rich Eureka-Idaho ore shoot was extending into its property. It also found itself in an awkward position, knowing that the capital necessary to sink a shaft another 1,400 feet (to intersect the pay shoot) was well beyond their means.

Sale of the Idaho Mine—1893

By the early spring of 1893 the Idaho mine had reached the compromise line agreed upon by the two companies in 1890. No other ore bodies had been discovered to sustain further operations. Therefore, the Coleman brothers felt it would not be easy to sell their mine to another company. The Maryland company would be the logical buyer because of the rich Idaho ore entering its property on three levels. The Coleman brothers had previously been on the Maryland board of directors, and they knew all to well that this company had little capital and was not in a position to purchase the Idaho mine. The Coleman brothers had made their fortunes, and retirement was looking very good to both of them. Edward Coleman, who had served as superintendent since 1875, resigned his position in April 1893. Eugene C. Creller was elected to fill his vacancy.

Negotiations soon followed between the two companies. An agreement was reached whereby the Idaho mine would be sold to the Maryland Gold Quartz Mining Co. for the sum of $85,000. The price was very reasonable, with easy payments.

The stipulations in the agreement were as follows: the Idaho Quartz Mining Co. agreed to sell, and the Maryland Gold Quartz Mining Co. agreed to buy, everything connected with the property known as the Idaho mine. This consisted of the hoisting works, shop buildings, mill, water rights, ditches, pipelines, reservoirs, and all land and mineral rights. All of the supplies on hand were also included. Excepted from the agreement was the money on hand, specimens in the vault, and gold in the mill. The two mines were to be worked as one under superintendent Eugene Creller, who represented the interest of the Idaho company. The interest of the Maryland company was represented by Victor L. Dorsey, youngest son of Samuel Dorsey. Mining operations were to continue as in the past, permitting the extension of the Eureka-Idaho-Maryland ore shoot to be mined in the Maryland property.

In regard to the payment for the property, the two companies agreed that 90 percent of the net monthly proceeds from the ore mined would be paid to the Idaho Quartz Mining Co. and applied to the $85,000 sale price until fully paid.

The dual management operation began on May 1, 1893. The mine's name was changed to Idaho Maryland mine, although the company name remained Maryland Gold Quartz Mining Co. The transaction was beneficial to the Maryland company, as it provided the opportunity to begin mining good ore from three levels within their own property. In addition, it provided the company with the means for making payments from the ore being mined. Otherwise the buyers would not have been able to purchase the Idaho mine, nor develop their own mine down to the depth of the ore shoot, due to lack of capital.

Local miners and townspeople were greatly relieved when the agreement was consummated. The community had serious concerns that the mine would close and result in a layoff of over 200 men. This would have had a great impact upon both Nevada City and Grass Valley.

Ore production under the dual management began in the Maryland mine on three previously mentioned levels. There had been no development work ahead in which to start stoping, nevertheless the drifts were run in good pay rock, which was hoisted as ore. The stopes were soon to follow, and the mill was able to run 30 stamps.

On May 12, 1893, a pocket of gold in large quantities was struck in the Maryland ground. Rich rock was also struck on the 1500-foot level which was unusually good. The ledge was very large on each level, and a number of fine specimens were extracted.

In the fall of 1893 all the surface area around the mine property available for dumping waste rock had been utilized, and it was necessary to provide additional space. Superintendent Creller conceived an idea that proved satisfactory. A hydraulic elevator was built about 50 feet east

A photograph of the Idaho Maryland taken by Victor L. Dorsey. Idaho Maryland road and the narrow gauge tracks are in the foreground.

Another Victor L. Dorsey photo, looking east to the Idaho Maryland mine, shows Wolf Creek behind teamster hauling cord wood.

of the headframe. The ram was 6½ inches in diameter and operated by high-pressure water. A single valve was used for both up and down movement. The structure was constructed of heavy framed timbers, anchored to a concrete foundation and extended high in the air. A car of waste rock was positioned on a platform at the waste tramway trestle and elevated 30 feet to a higher trestle. This provided a considerable amount of additional area in which to dump waste rock.

A Westinghouse dynamo with a 60-light capacity was installed on the surface. It provided lighting for all buildings and specific yard areas. The underground activity was well organized by the end of the year. The mine was producing good ore. The production amounted to $258,220, from which monthly payments were made toward the mine purchase.

Near Catastrophe—1894

In January 1894, when full management was about to be turned over to the Maryland company, the hoisting works and other equipment were severely damaged by fire, causing a work stoppage and allowing the lower mine workings to fill with water. This large conflagration ignited at approximately 9:30 p.m. on the evening of January 9, 1894—a fire that could have been the worst in the history of this district.

The fire started from a defective flue in the dry building. The flames spread so rapidly that nothing could be done to stop their progress. The shaft collar, headframe, air compressors, hoists and pumping equipment were all housed within a very large wooden structure. The blacksmith shop was also attached. Within a short time the entire complex was engulfed in flames.

Immediately upon noticing the flames, the hoistman blew the mine whistle to summon the Grass Valley volunteer fire department. Townspeople knew all too well the meaning when hearing a mine whistle at unusual hours, and they were quick to investigate. They saw a bright red glow, visible throughout Grass Valley. Hundreds of people quickly lined the Nevada County Narrow Gauge railroad tracks across from the mine, awaiting word of the fate of the men working underground.

The night shift, comprising 61 miners, was underground when the fire occurred. The shaft collar, from which the miners had descended about six hours earlier, was completely surrounded by intense heat, flame and poisonous gasses. Also, the normal flow of fresh air for the underground workings was through the shaft. The air continued downward, making its way throughout the workings, before exhausting through the air shaft, 600 feet to the east. The immediate concern was to quickly notify the miners in order that they might escape before the smoke and poisonous gasses reached them in amounts that would endanger their lives.

Arriving at the mine from their homes, superintendent Creller and Victor Dorsey took charge of forming search parties that would get word to the men in the mine and assist them to safety. At 10:00 p.m., a party consisting of Ben Opie, Si Rowe and Alfred Argall went down the vertical pump shaft. Their efforts proved futile, and they returned to the surface. Another volunteer search party composed of Edward McGuire, Will Corbett, Frank Bawden, Isaiah Moyle, William Richards and William R. Thomas descended the vertical air shaft.

Meanwhile, shortly after 9:30 p.m., Tom Rodgers, Canyon shaft hoistman on the 1000-foot level, was first of the underground crew to know something was wrong. He suddenly lost air pressure to his hoist. He and Joe McGrath then walked to the main shaft, which was only 300 feet away. There was no response when they rang for a cage, which was spotted just above the station. This was their first indication of trouble in the mine. While they were standing on the station one of the flat hoisting cables came falling down the shaft. In moments it was followed by the "dolly cage," which was full of fire.

Tom Rodgers and Joe McGrath immediately realized the seriousness of the situation and proceeded down the Canyon shaft to warn the men on each of the lower levels. After all men were notified, the long strenuous climb out of the mine began. Shift boss Forrest Farnham was familiar with the exit route, and he led the men.

The first 600 feet was by wooden steps up the 40° Canyon shaft to the 1000-foot level. From there to the 400-foot level the avenue of escape was through six separate raises through old workings. Each of these raises was 70° or greater, and each was 100 feet in length with step-ladder ways. Egress from the 400-foot level to the surface was by ladder through the vertical air shaft.

Scores of weeping parents, wives, children and other relatives waited anxiously near the air shaft collar, praying for the safe return of the miners. At 11:00 p.m. members of the search party returned to the surface with word that the miners were all accounted for and were on their way out.

The first to reach the surface was James Ledwich. Everyone closed in around him to learn about the rest of the miners. Tears of joy trickled

Miners Underground When the Fire Started in 1894

Bastian, Jack	Hogan, James	Paul, Charles
Bastian, James	Hooper, Harry	Paul, William
Bone, William	Hurley, Jerry	Peterson, P. F.
Bracy, Edward	Jaecks, Robert	Polkinghorn, Jack
Buckley, Mike	James, James	Polkinghorn, Richard
Burns, Buts	Jenkins, Alfred	Richards, John
Burns, William	Jennings, Thomas	Rodgers, Thomas
Carter, George	Jones, Frank	Ryan, Richard
Carter, Richard	Jones, Robert Jr.	Shoemaker, Harry
Cole, William	Juliff, James	Stevens, William
Conley, Richard	Ledwich, James	Stevens, William H.
Coombs, Jacob	Madden, William	Temby, William Jr.
Cornish, William	May, T.	Thomas, Charles
Davidson, James	McCurry, William	Thompson, George
Dixon, Joseph	McGrath, Joseph	Trezise, James
Farnham, Forrest	McGregor, Charles	Tribilcox, John
Gill, Frank	Mitchell, Thomas	Uren, Stephen
Hall, Thomas	Moyle, William	Wasley, Samuel
Hayes, James	Painter, Thomas	Williams, John K.
Hicks, Richard	Partridge, Samuel	Williams, Thomas
	Pascoe, Edwin	

down the faces of those present, as one by one the miners reached the surface.

Shift boss Forrest Farnham had waited until all men had gone ahead before he ascended the ladder. Each man was fatigued as a result of the long arduous climb. No one was injured, although several confessed that they had been frightened upon first hearing the news of the fire.

The large mill building was saved through the heroic efforts of the firemen and many townspeople who worked to suppress the fire. It was necessary to shut down the Pelton wheel that operated the dynamo, in order to conserve water for fire fighting. By 1:30 a.m. all that was left of much of the surface works was a smoldering mass of ruins.

Damage Assessment—1894

The next morning the damage was assessed and the loss was estimated to be $100,000. The insurance coverage for the entire surface works amounted to only $32,000. The headframe and all hoisting equipment were completely destroyed. The large Cornish pump received major damage, but was repairable. The shaft timbers had burned down the shaft for a distance of 20 feet. It was fortunate that the fire had burned so hot, and atmospheric conditions were just right for the smoke to rise and not be drawn down the shaft. The dry building and blacksmith shop were completely destroyed.

Management's main concern was how much of the mine would fill with water before pumping could resume. The cost to pump from the lower levels would be high if these workings were to become flooded. The smaller Cornish pumps in the vertical pump shaft were unharmed and continued raising water from the 1000-foot level and above at a steady four strokes per minute.

Superintendent Creller announced that the removal of the debris would begin immediately, in preparation for the erection of new buildings and installing new equipment. Priority was given to repairing the Cornish pump, in order to begin removing water from the lower levels. The timbers were replaced at the shaft collar and for several sets down the shaft.

Beginning of the Dorsey Era—1894

The Maryland company took full possession of the Idaho Maryland mine on February 1, 1894. The following management changes were made: Samuel Dorsey, president and superintendent; Victor Dorsey, assistant superintendent; Forrest Farnham, foreman; and John Carter, head timberman. Eugene Creller, who had been superintendent during the transition period, tendered his resignation.

To guard against a catastrophic fire in the future, Samuel Dorsey had a series of water standpipes installed. These were equipped with hoses, and appropriately positioned near each building. A hydraulic monitor was also installed on the hill above the mill and was connected to a high-pressure line. This "little giant" was capable of deluging all buildings with a tremendous stream of water.

Just as soon as the new hoisting equipment and Cornish pump became operational, the flooded levels and stopes, down to and including the 1600-foot level were pumped free of water. The water in the Canyon shaft was pumped down a short distance below the 1600-foot level, and the water was held at that point.

Mining resumed as quickly as possible on the three levels where it had been interrupted. Some good ore was found that lasted throughout the year. The Maryland company wasn't as fortunate as its two predecessors, for the Eureka-Idaho-Maryland ore shoot extended only about 200 feet into the Maryland beyond the compromise line. The vein narrowed after leaving the ore shoot and became unprofitable to mine.

The gold production for the partial year of 1894 was $193,182. The employment was considerably less than the previous years, with only 90 men underground and 23 on the surface.

Dorsey Vein

Just when the main vein was about to give out, the Dorsey vein was discovered in the footwall. The quartz appeared very strong, with good ribboned sulfides and a fairly good grade of ore.

At a point approximately 2,000 feet east of the Canyon shaft on the 1600-foot level, a winze was started on the new vein. The Dorsey winze proved to be another link to the already complicated underground movement of ore. This involved maintaining three hoisting works, two of which were underground. It also meant tramming the ore long distances.

Victor Dorsey Fatality—1895

A distressing accident occurred on November 5, 1895, through which death invaded the family of Samuel Dorsey, principal owner of the mine. His son Victor had gone with foreman Forrest Farnham and specimen boss William McCurry to the bottom of the Dorsey winze, below the 1600-foot level. They had just finished their inspection of the vein, and all but Victor had moved away from the face of the incline; as he started to walk away, some rock fell from the hanging wall, striking him on the back.

Farnham, McCurry and two miners quickly removed the rock and brought the injured man to the surface. Dorsey was treated by his brother-in-law, Dr. W. I. Howard, before being placed on a mattress in a horse-drawn express wagon and transported to his home. He died shortly after, from crushing injuries to his chest and rib cage. He was also treated by Dr. George O. Willis, but the two physicians were unable to save his life.

During the year 1895 raises and stoping continued on the Dorsey vein and also on remnants of the main vein on the 1500- and 1600-foot levels. Sinking continued in the Dorsey winze to explore the vein it its downward dip. The total production for the year amounted to $247,600.

Year 1896

Going into the year 1896 the company was faced with considerable

The Greatest Event

Of the Season.

GRASS VALLEY

Miners' Union Picnic.

WILL BE HELD AT

Ismert's Grove and Glenbrook Park,

JUNE 5, 1895.

Double Handed Drilling, free for all—First prize, $40; second, $20.

Single Handed Drilling, free for all—First prize, $20; second, $10.

Also, a Single Handed Drilling Contest for Members of the Union Only—First prize, silver medal and one year's subscription to the Morning Union; second, one year's subscription to the Daily Tidings and 1 box of Cigars.

Additional Prize for Double Handed Drilling—Two Silk Finished Vests by J. P. Miller.

Baseball Match—Prize, $20 and 5-gallon Keg of Beer.

Football Match—Prize, $20.

Throwing the Hammer, Foot Races, Wheelbarrow Races and numerous other games too numerous to mention.

☞ Entrance Fee: Double-Handed Drilling, $5.00; Single-Handed Drilling, $2.50. Entries close June 3d.

Trains will run every hour of the day and to and from picnic grounds. Timetable will be published later. Fare for round trip and admission to grounds, 50 cents. Children under ten years of age, half price.

Round trip from Colfax and admission to grounds, $1.

Admission to grounds other than by rail, 25 cents.

DANCING—Gentlemen, 50 cents; ladies, free.

The Maryland Gold Quartz Mining Company

CAPITAL STOCK SHARES.	*PAR VALUE PER SHARE:*
500,000.	$20.00.

GRASS VALLEY MINING DISTRICT,

NEVADA COUNTY, CALIFORNIA.

Map showing the direction of the famous

EUREKA—IDAHO—MARYLAND LEDGE.

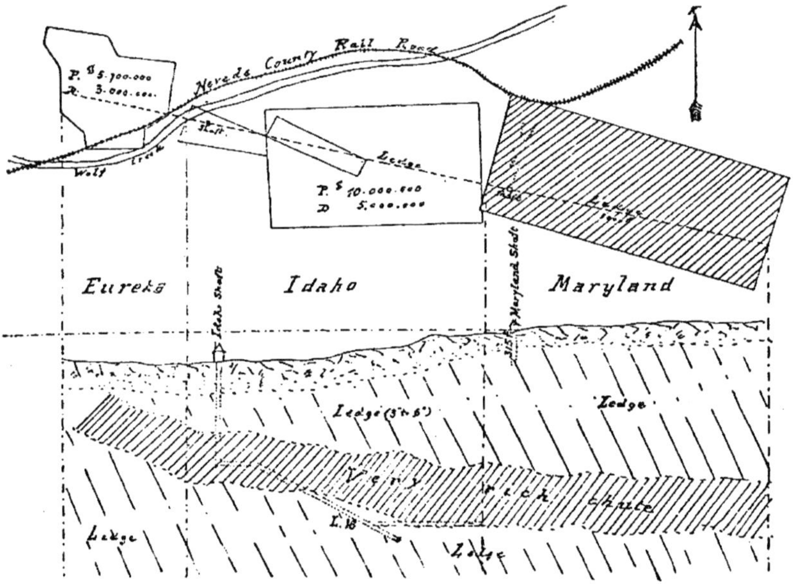

Scale—20 chs. to the inch.

LONGITUDINAL SECTION OF THE VEIN.

FOR SALE.

40,000 SHARES AT $2.00 A SHARE

TO BE USED EXCLUSIVELY FOR THE DEVELOPMENT OF THE PROPERTY.

For further particulars and share subscription inquire of

S. P. DORSEY, Express Office, Grass Valley, Nevada Co.

timber replacement problems. This was due in part to the Idaho company's neglecting timber replacement in several areas, knowing that the mine would be sold or closed. More importantly, though, timber replacement had to take place on an ongoing basis, not only in the drifts but in the shafts as well. Serpentine in the footwall formation swelled when exposed to moisture in the air, and this placed constant pressure against the timber, causing it to break.

The Maryland company lacked sufficient funds to maintain the underground workings in a first-class condition. Timber repair and replacement was treated on a priority basis. The high cost of underground maintenance work continued to plague the company throughout its operation.

The Dorsey winze continued down to a depth of 450 feet along the dip of the vein, with three levels established from the winze. The 1700-foot level was driven 360 feet, the 1800-foot level 115 feet, and the 1900-foot level 250 feet. The vein on the 1900-foot level ranged from two to four feet in width.

The work of driving drifts and sinking the winze was done on a day's pay basis by the company. In the latter part of 1896 much of the stoping operations was performed by tributers. The total gold production for 1896 was $197,239.

Years 1897–1901

The major problem confronting the Maryland company was insufficient capital to maintain the underground workings, and at the same time develop new ore bodies. It was necessary for a mine to maintain at least one—and preferably two—year's worth of ore reserves ahead at all times. Samuel Dorsey offered to sell 40,000 shares at $2 a share. The money was to be used exclusively for development of the underground veins. The money from outside interests failed to materialize.

Mining continued for the next four years, with production declining each successive year. During the four-year period gold production totaled $259,775. Included in this figure was $20,546 contributed by tributers from 2,708 tons of ore milled between November 1896 and August 1899. The tributers' rock ranged from $5 to $30 per ton, with an average of $8.49 per ton.

A few days before the mine closed on February 1901, a group of prominent mining men was asked to conduct an inspection of the lower workings of the mine. The request was made by Louis Dorsey, son of Samuel.

Those involved in the inspection were Charles Stocks, formerly foreman of the North Star mine; William F. Englebright, mining engineer and superintendent of the South Yuba Water Co.; Charles A. Brockington, superintendent of WYOD (Work Your Own Diggings) mine; George W. Starr, general manager, Empire mine; Robert Walker, superintendent, Gold Hill Mining Co.; and Samuel Alderman, mining and civil engineer.

A report signed by the men named above was given to Louis Dorsey on February 14, 1901. The following is quoted from their sworn statement:

> "We, the undersigned, have made a personal inspection of the Idaho Maryland Mine, Grass Valley Mining District, and in our judgement the ledge we saw in the east incline or Dorsey shaft, and the levels from the shaft, is a continuation of the Eureka-Idaho-Maryland pay shoot of quartz, on its eastward and downward strike, in the regular fissure, and of good size, viz, two to four feet, and appears to be of good quality."

In February 1901 the Idaho Maryland mine closed and was allowed to fill with water. There were several immediate causes for the closing: lack of capital to develop the mine, cost of triple handling of ore, high milling costs owing to sporadic flow of ore to the mill, and poor condition of underground workings. This ended 44 years of continuous mining of gold at this location.

A Phoenix Rock Drill in 1894 cost as little as $100 for a 2" model to as much as $235 for a 3⅝" drill. It was guaranteed "to drill more feet of holes for the same money expended in labor and repairs than any other machine."

CHAPTER 5: IDAHO MARYLAND DEVELOPMENT COMPANY

Following the mine closure in 1901, Samuel P. Dorsey kept the surface workings intact. He became active in promoting the property, in hopes of enticing sufficient capital to reopen the mine. In the latter part of 1902 a group of men from Boston became interested, and they sent a mining engineer to Grass Valley to review the property, underground maps, previous reports, and the surface works. The engineer compiled a thorough report, which proved satisfactory to the prospective investors.

The Idaho Maryland Development Co. was thus formed, with its main business office at 101 Milk Street, Boston, Massachusetts. The company was capitalized at $3,000,000, par value of the stock being one dollar per share. There were 1,500,000 shares in the treasury.

The following officers were elected: Thomas W. Shapleigh, president, was with the American Sugar Refinery Co., also treasurer of the Boston and Portland Steamship Co.; E. M. Preston, vice president, was a judge and prominent businessman; George E. Marvin, second vice president, was president of the Loder Pyrite Reduction Co.; and Henry P. Colsson, secretary and treasurer, was treasurer of the American Steamboat and Hotel Co. of Boston.

The four named men were directors, as was John Carter, a wholesale paper manufacturer. Alphonse A. Wyman was the company's counsel. Cornelius B. Lakenan, mining engineer of Grass Valley and superintendent of the Coe mine, was hired to head up the engineering department.

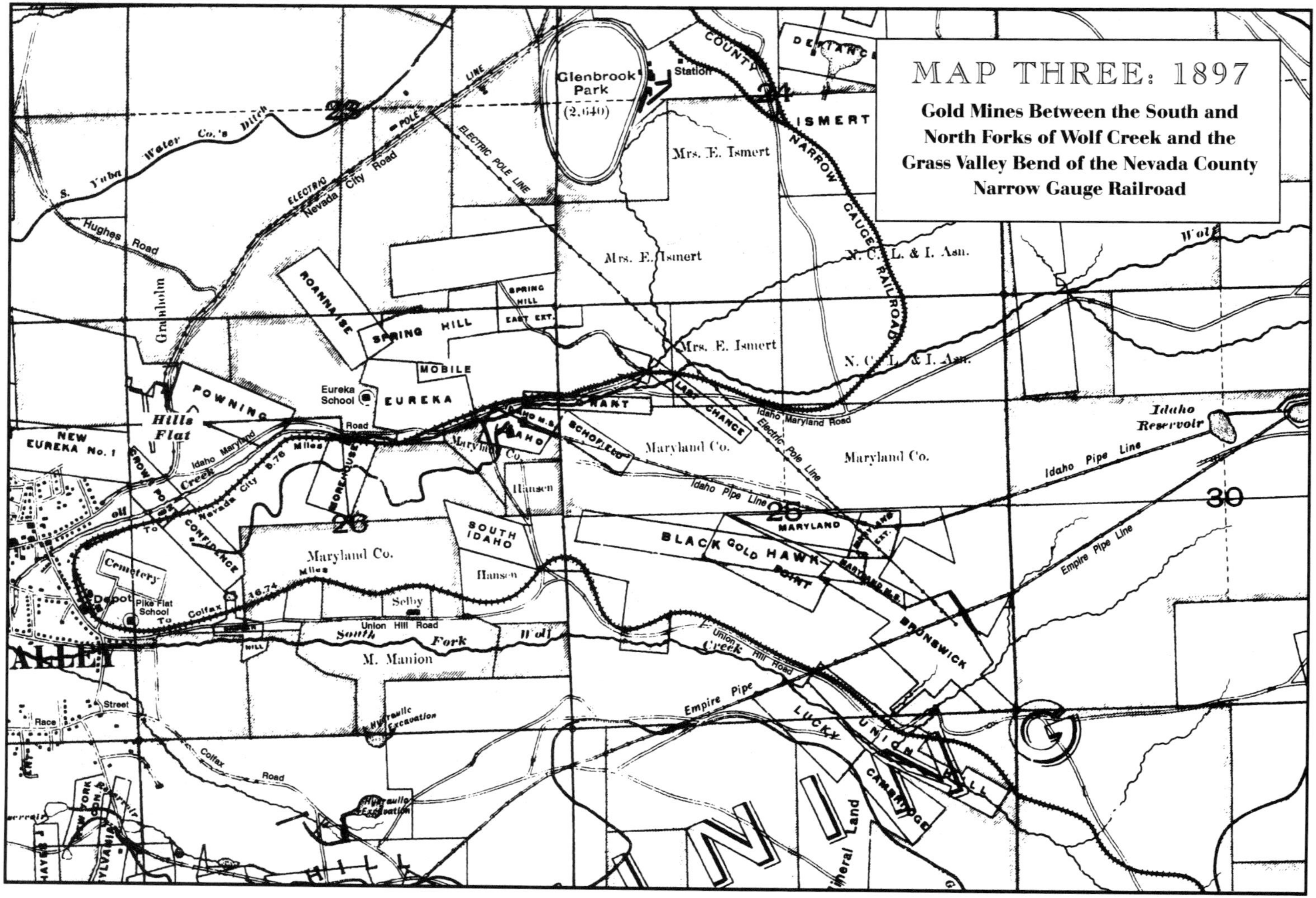

MAP THREE: 1897

Gold Mines Between the South and North Forks of Wolf Creek and the Grass Valley Bend of the Nevada County Narrow Gauge Railroad

The International Trust Co. of Boston was named as depository, while Wells, Fargo Company's Bank of San Francisco was assigned to receive all bullion and act as trustee for the stockholders.

In 1903 the Idaho Maryland mine was bonded to the Idaho Maryland Development Co. This new company looked forward to reopening the mine with great enthusiasm. Several suggested plans had been advanced for developing the extensions of the known ore bodies. Two recommended plans involved sinking new shafts in hard ground 2,000 feet east of the Idaho shaft. These were rejected due to cost considerations.

The plan decided upon was to reopen the Idaho shaft to the 1000-foot level, and then continue sinking the shaft in the unexplored portion of the vein below the Eureka-Idaho-Maryland ore shoot. Levels would then be run as would be warranted by the development. The shaft would terminate at the 2000-foot level, where a drift would be run 1,500 feet east to encounter the Canyon shaft ore shoot, then continue drifting further east a distance of 2,000 feet to the bottom of the Dorsey winze.

This would permit the mine to be worked from one shaft, thus eliminating the triple handling of ore through the Dorsey winze and Canyon shaft. It would also reduce the pumping cost by a more simple method of pumping through only one shaft.

The enthusiasm shown by the Boston company was quickly dimmed when actual work was started to reopen the mine. The overall condition of the surface and underground workings had deteriorated greatly under the Dorsey regime, due to the lack of capital to maintain the mine in first-class condition. Among the first things to confront the company were major equipment repairs needed in the mill, hoisting works and the pumping system. Then, in July 1904, the surface repairs were completed and pumping was commenced to dewater the underground workings.

As the water level receded in both the Idaho shaft and the vertical pump shaft, it was immediately evident that the shaft timbers were in bad condition and in dire need of replacement. The unexpected high cost to retimber the two shafts, and the time consumed to accomplish the work, caused the company to alter its original plan to continue sinking the Idaho shaft. The company decided instead to reopen the upper mine levels in hopes of finding enough ore to sustain operations while shaft repairs were underway.

Much of the recapture work of the old drifts and ore extraction was

JOHN MARTIN, President
31-33 New Montgomery St., San Francisco

Dressed Flooring and Siding
Best Sugar Pine
Mining Timbers a Specialty

Building Lumber
constantly on hand

Shakes, Shingles and Laths
Sash, Doors, Windows
Lumber Sawed to Order on Short Notice

WM. B. MEEK, Sec'y and Treasurer
Camptonville, Cal.

Grass Valley, Cal., July 1904

M Idaho Maryland Development Co.

To **Yuba Development Company**, Dr.

Wholesale and Retail ..... LUMBER DEALERS

AL. H. SMITH, Manager

Terms ______ Interest after Maturity ... AUBURN STREET.

Date	Item		Feet	Total Feet	Rate	Amount	Total
July 7	20-2-12-16	Com	640				
	4-8-8-12	"	256	896	18	16 13	
9	6-10-12-12	"	720				
	25-1-12-16	"	400				
	2-6-12-16	"	192				
	1-6-12-18	"	108	1420	"	25 56	
13	2-18-24-40			2880	30	86 40	
	6-16-18-12			1728	18	31 10	
19	3-2-12-12		72				
	10-2-12-16		320				
	10-2-12-18		360				
	6-6-6-18		324	1076	"	19 37	
20	50-1½-12-16			1200	"	21 60	
21	35-1-12-16		560				
	3-4-6-16		96				
	3-4-4-16		64				
	10-6-6-18		540				
	6-1-6-16		48				
	4-12-12-12		576				
	2-10-12-16		320	2204	"	39 67	
22	12-2-12-18		432				
	20-2-12-16		640	1072	"	19 30	
23	1-4-6-20		40				
	18-6-6-18		972				
	15-1½-4-16		120				
	20-2-4-16		213	1345	"	24 21	
27	16-1½-18-16		576				
	6-1½-14-16		168	744	"	13 40	
28	35-1½-16-18		1260				
	4-1½-14-16		112				
	40-½-4-16		213				
	56-1-12-16		896				
	5-1½-16-18		180				
	1-15-15-16 2/		309				
	1-15-15-22 2/		422	3392	"	61 06	
30	4-15-15-22 2/		1688				
	1-15-15-16½		309				
	26-1-12-16		416				
	1-15-15-16 2/		309				
	2-10-15-19		475	3197	"	57 55	415 35
				2%			8 31
							407 04

Paid Aug 4-1904
Yuba Development Co
By [illegible]

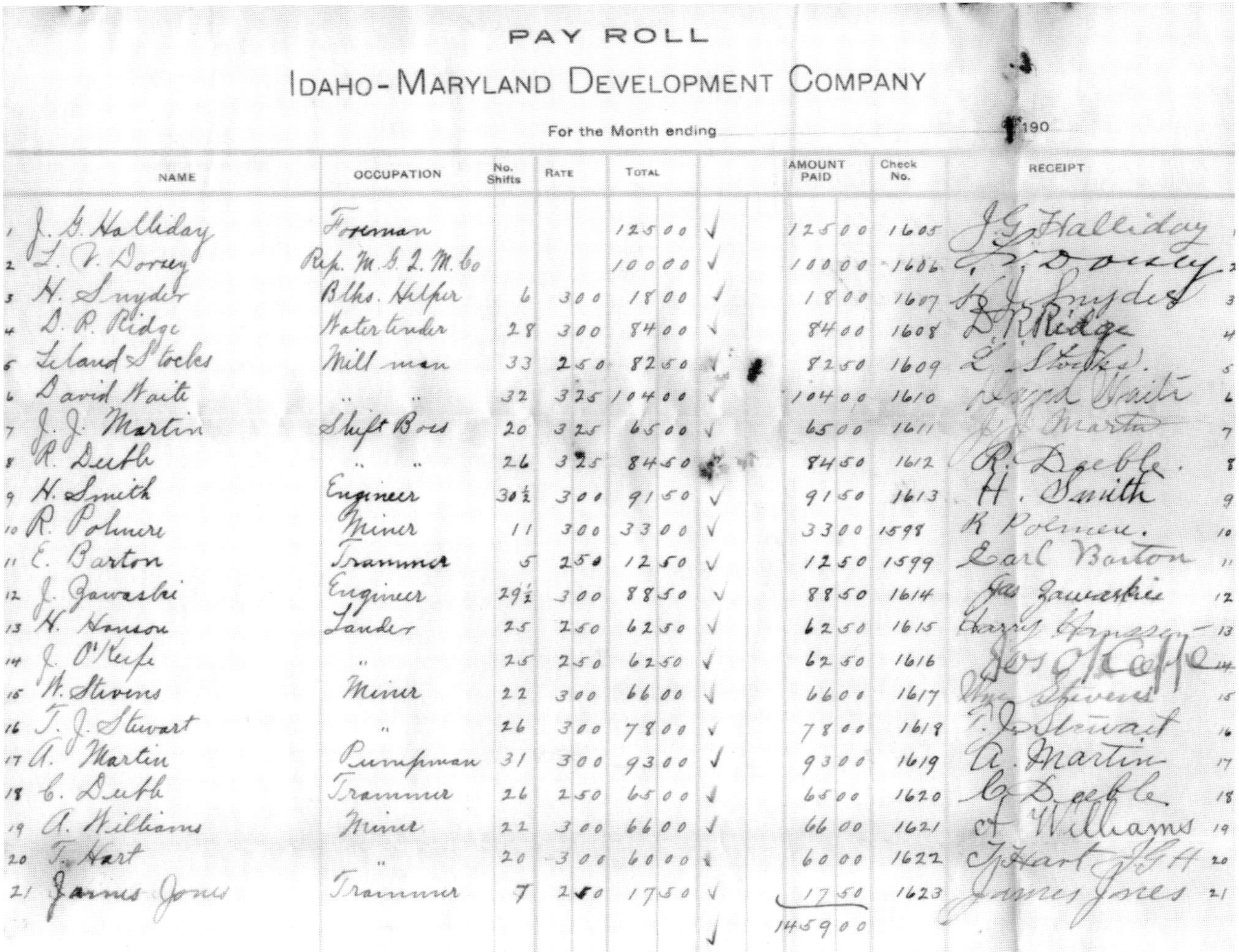

PAY ROLL

IDAHO-MARYLAND DEVELOPMENT COMPANY

For the Month ending ________ 190

	NAME	OCCUPATION	No. Shifts	RATE	TOTAL		AMOUNT PAID	Check No.	RECEIPT	
1	J. G. Halliday	Foreman			125 00	✓	125 00	1605	J. G. Halliday	1
2	L. V. Dorsey	Rep. M. S. I. M. Co			100 00	✓	100 00	1606	L. V. Dorsey	2
3	H. Snyder	Blks. Helper	6	3 00	18 00	✓	18 00	1607	H. J. Snyder	3
4	D. R. Ridge	Water tender	28	3 00	84 00	✓	84 00	1608	D. R. Ridge	4
5	Leland Stocks	Mill man	33	2 50	82 50	✓	82 50	1609	L. Stocks	5
6	David Waite	" "	32	3 25	104 00	✓	104 00	1610	David Waite	6
7	J. J. Martin	Shift Boss	20	3 25	65 00	✓	65 00	1611	J. J. Martin	7
8	R. Deeble	" "	26	3 25	84 50	✓	84 50	1612	R. Deeble	8
9	H. Smith	Engineer	30½	3 00	91 50	✓	91 50	1613	H. Smith	9
10	R. Polmere	Miner	11	3 00	33 00	✓	33 00	1598	R. Polmere	10
11	E. Barton	Trammer	5	2 50	12 50	✓	12 50	1599	Earl Barton	11
12	J. Jawaski	Engineer	29½	3 00	88 50	✓	88 50	1614	Jas Zawaskie	12
13	H. Hanson	Lander	25	2 50	62 50	✓	62 50	1615	Charles Hanson	13
14	J. O'Keefe	"	25	2 50	62 50	✓	62 50	1616	Jos O'Keefe	14
15	W. Stevens	Miner	22	3 00	66 00	✓	66 00	1617	Wm Stevens	15
16	T. J. Stewart	"	26	3 00	78 00	✓	78 00	1618	T. J. Stewart	16
17	A. Martin	Pumpman	31	3 00	93 00	✓	93 00	1619	A. Martin	17
18	C. Deeble	Trammer	26	2 50	65 00	✓	65 00	1620	C. Deeble	18
19	A. Williams	Miner	22	3 00	66 00	✓	66 00	1621	A. Williams	19
20	T. Hart	"	20	3 00	60 00	✓	60 00	1622	T. Hart J. G. H.	20
21	James Jones	Trammer	7	2 50	17 50	✓	17 50	1623	James Jones	21
						✓	1459 00			

As it neared the end of its operations, the Idaho Maryland Development Co. had very few men left on its payroll, as this ledger sheet demonstrates.

done by tributers. The shaft timbering was done on a day's pay basis by company men. These upper workings dated back 20 to 25 years, when this part of the mine had been worked under the Coleman brothers. All of the timbers were crushed from swelling ground and deterioration from age. The recapture of the levels was slow, and proved quite costly, even by use of tributers.

Many of the old stopes were found to be backfilled with gob, which had values between $2 and $4 per ton. For the next several years mining consisted mostly of tributers recapturing the old drifts, removing the gob and fines from old stopes (where assays warranted) and removing pillars left by previous operators. The milling costs were much higher than normal because of the sporadic flow of ore from underground.

New Ore Discovered—1908

In 1908 a vein 18 inches in width, and lying parallel to the main Idaho vein on the 500-foot level, was found to have good values. A drift was advanced 200 feet, where the quartz had widened to five feet. The ore from drifting and stoping provided sufficient mill rock to run ten stamps. The same vein was later intersected on the 700-foot level, where the quartz was eight feet wide. General manager Bray Wilkins granted several blocks of ground to different groups of tributers, bringing the underground force to 75 miners. This ore body extended from the 700-foot level to above the 400-foot level, which allowed the mill to operate 30 stamps well into the year 1911.

During the period of operation under the Idaho Maryland Development Co. the 400-foot level was reopened for 650 feet, and the 500-foot level was reopened 1,500 feet to the east. The 600-foot level was reopened to the pump shaft, and the 700-foot level was reopened to the west property line, and 1,320 feet to the east. The 800-foot level was reopened 100 feet to the east and 80 feet to the west. The 900-foot level was reopened 500 feet east, and a new drift was run 115 feet to the west. The main shaft was retimbered, and the water kept pumped down to the 1000-foot level, but no effort was made to advance that level, nor to dewater the Canyon shaft.

One thing was for certain: the Idaho Quartz Mining Co. did not overlook much ore in their previous workings, nor did the development company find any significant new vein structures. The Idaho Maryland Development Co. ceased operations in October 1914, at the beginning of World War I. The mine again was allowed to fill with water. From

1904 until October 1914, the Idaho Maryland Development Co. had a total production of $236,613 for the company account, to which $75,000 should be added for the tribute account.

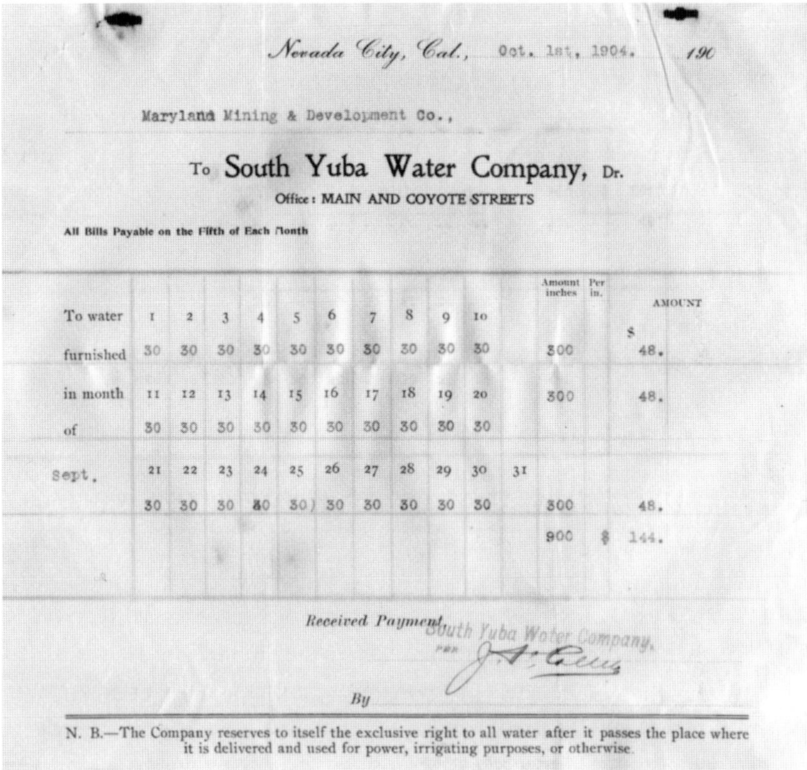

Nevada City, Cal., Oct. 1st, 1904. 190

Maryland Mining & Development Co.,

To **South Yuba Water Company,** Dr.

Office: MAIN AND COYOTE STREETS

All Bills Payable on the Fifth of Each Month

To water	1	2	3	4	5	6	7	8	9	10		Amount inches	Per in.	AMOUNT
furnished	30	30	30	30	30	30	30	30	30	30		300		$ 48.
in month	11	12	13	14	15	16	17	18	19	20		300		48.
of	30	30	30	30	30	30	30	30	30	30				
Sept.	21	22	23	24	25	26	27	28	29	30	31			
	30	30	30	30	30	30	30	30	30	30		300		48.
												900		$ 144.

Received Payment, South Yuba Water Company, per J. A. Ellis

By

N. B.—The Company reserves to itself the exclusive right to all water after it passes the place where it is delivered and used for power, irrigating purposes, or otherwise.

Grass Valley, Cal., May 1 1906

M Idaho Maryland Dev Co

TO **FASHION STABLES,** DR.

W. G. LORD, PROPRIETOR

Horses Boarded and Given Careful Attention. *Single and Double Teams.*

Hacks for Any Occasion.

STABLES, LOWER MAIN STREET

TELEPHONE, BLACK 473

1906

Apr 9 To Team [illegible] 6

30 Boarding Horse [illegible] 22 50

[illegible] 28 50

CHAPTER 6: CONSOLIDATION BY ERROL MACBOYLE

In 1914 the once prosperous Idaho Maryland mine lay idle with all of its underground workings full of water. Later that year, Errol MacBoyle, a well-known mining engineer from San Francisco, became interested in the Idaho Maryland mine. He had made a comprehensive study of the mine's records, geology and history while serving as a reporting engineer for the California State Mining Bureau. His study of the mine convinced him that this mine contained vastly greater riches than the 19 million dollars it had yielded to previous owners.

Fortified with this knowledge and his faith in the mine, MacBoyle set out with limited resources to acquire an option on the mine. He also envisioned the great advantages of consolidating all of the surrounding mines together with the Idaho Maryland. There was sound engineering in this idea, for veins run into adjacent properties and cannot be followed. That was why the Empire and North Star companies of Grass Valley had for years a standing offer to purchase the properties contiguous to their boundaries.

Errol MacBoyle then interested two other well-known engineers to assist him. One was Frederick W. McNear, who was president of the Union Hill mine at Grass Valley, where MacBoyle served as a consulting engineer. The other was Edwin Letts Oliver, president of Oliver Filter Co., with whom MacBoyle had worked years earlier at the North Star mine. The three engineers had long been familiar with the workings of the Empire and North Star mines, and could not believe the Idaho Maryland would be without ore bodies below the 2000-foot level to which it had been worked. They pointed out that the Empire and North Star had found profitable ore in much greater depths.

MacBoyle received his first option, from the South Idaho Consolidated Mining Co., on June 1, 1915. The second option was granted on July 3, 1915, from the Gold Point Gold Mining Co. The third option was granted by Frederick Snook, consisting of the Black Hawk mine and several adjacent mining claims.

To further his plan, MacBoyle formed a syndicate to acquire and develop the Union Hill mine, the members of which were himself, McNear, Oliver, Snook, and Roy H. Elliott. The Union Hill mine was in operation at that time, with Frederick McNear as president. Later, Gold Point Consolidated Mines Inc. was formed to take over the interests of the syndicate and purchase additional properties.

The main thrust of the organization was to reopen the Idaho Maryland mine, but financing a venture of that magnitude was well beyond the ability of the new corporation. It was wartime and raising capital was very hard to achieve.

During the year 1917 the Metals Exploration Co., with offices at 120 Broadway, New York, acquired a controlling interest in Gold Point Consolidated Mines Incorporated. Harry Payne Whitney was president of Metals Exploration, and Bulkely Wells was vice president. Wells was elected president of Gold Point, replacing Frederick McNear. The new directorate included Bulkely Wells, Rufus Thayer, and John Crossen, in addition to MacBoyle, McNear, Oliver and Elliott.

After many months of negotiations, one by one, properties were leased with options to buy. These included the Eureka, Roanaise, South Idaho, Black Hawk, Union Hill, Gold Point, and Idaho Maryland mines, and several other mining claims. The land involved extended over two miles in length on the Eureka-Idaho-Maryland lode, and embraced an extensive vein system, the greater part of which was in virgin ground. The last property to be added was the Idaho Maryland, which was purchased on December 20, 1919, from the estate of Samuel P. Dorsey and the Dorsey family for $187,692.50. By this time, in excess of one million dollars had been expended in acquisition of the entire holdings.

After careful study, a plan evolved for the development of the combined properties, at an estimated cost of another one million dollars. In order to secure this amount of capital, an agreement was entered into in 1919 with the Metals Exploration Co. by which it acquired controlling interest in the combined properties and advanced the funds necessary to equip, reopen, develop, and bring the project to a self-supporting basis. In return, they received bonds and notes equal in amount to the money advanced by them. In addition, one share of stock was given as a bonus for each dollar invested.

The name of the new venture was Idaho Maryland Mines Co. The officers and directors were Bulkely Wells, president; Roy H. Elliott, vice president; and Errol MacBoyle, consulting engineer. Frederick McNear, Rufus Thayer and Robert Anderson were also directors. John A. Fulton was appointed general manager.

The Union Hill mine, showing mill and waste dumps.

CHAPTER 7: REOPENING IDAHO MARYLAND MINE

The Metals Exploration Co. intended to sink a new vertical shaft several thousand feet east of the Idaho shaft when they took control of the consolidated group of mines. This shaft would be sunk in hard rock, and bottom close to the Dorsey winze workings, where good ore was known to exist. It would be in a central location, from which the Union Hill, Gold Point, Black Hawk and South Idaho mines could be developed.

A close review of the Idaho Maryland showed the surface plant to be very antiquated, dating back to the Coleman brothers regime. The hoisting equipment had been operated by water power and equipped with flat steel ropes. The ore had been hoisted, one tram wagon at a time, on specifically designed open cages. Cornish pumps had been used for pumping mine water, and these needed considerable repair. Therefore, one of the first projects undertaken was to dismantle the entire surface plant.

Everything was removed from the mine site, with the exception of the office and dry building. This seemed to be the right decision at the time, because the new centralized shaft was in the planning stage. The new shaft plan was abandoned, however, in favor of one to reopen the mine through the Idaho shaft, and sink the shaft deeper to explore the Eureka-Idaho-Maryland vein below the 1000-foot level.

Miners Strike—1919

When the Metals Exploration Co. took over the consolidated group of mines, the Union Hill was the only mine in operation. An accelerated development program had been put in place that included drifting on the Lucky Cambridge vein on the 600-foot level, drifting on the Gold Point vein on the 800-foot level, and crosscutting on the 300-foot level to intersect the Georgia vein.

Shaft pockets were installed for the 1200-foot level and a crosscut was extended. Short drifts were started on four veins cut by the crosscut. New rock drills were purchased in preparation for working the new faces.

The underground work came to a halt on June 13, 1919, when the miners of the district called a general strike of all mines. Two weeks had passed before the strike was settled, and by that time water in the Union Hill mine had risen to the 600-foot level.

The new development work had not proven encouraging, therefore a decision was made to allow the mine to continue to fill with water. Operations were permanently suspended, and all employees terminated. The Metals Exploration Co. had expended $45,797 on development work within a five-month period with no return.

The strike was settled by the mining companies agreeing to grant a 10 percent wage increase to every employee who had worked three months, to pay one-half the time the men were out on strike, and to establish a butcher shop to sell meat to miners at a reduced price. This settlement became effective on June 30, and the agreement was for 12 months. The new daily wage scale was: miners $4.40, muckers $3.85, laborers $3.85, and shaft men $4.85.

Rebuilding the Surface Plant—1919

The first priority following the decision to reopen the Idaho shaft was to rebuild the surface plant. Fortunately, the Union Hill plant had become available, and that mine's entire surface works was dismantled and moved to the Idaho Maryland mine site. A changeover from water power to electricity was made at that time. A bank of three Westinghouse transformers, 4,000 volts, 200 kva capacity each, were installed on the surface. This was ample electricity to supply both the surface and underground power needs.

The 74-foot wooden headframe was erected at the shaft collar, and a new 75-ton capacity ore and waste bin constructed in the upper portion to accommodate self-dumping skips.

The Union Hill hoist was transformed from waterpower to an electric drive with herringbone gears, and a 150 hp variable-speed motor was added. The drums were modified to provide greater braking capacity.

The hoist was capable of hoisting 2-ton skips, with a rope speed of 750 feet per minute. The Sullivan and Ingersoll Rand air compressors from Union Hill were set in place. Larger pulleys were installed on each compressor, increasing the speed of each, giving a combined capacity of about 2,200 cubic feet of free air per minute.

Shaft Dewatering—1919

After the headframe was completed, the hoist set in place, and the air compressors operational, attention was directed toward dewatering the mine. First, the collar of the shaft had to be retimbered, and the drain tunnel, 23 feet below the collar, was reopened and retimbered for a distance of 360 feet.

An airlift method was devised to remove the water. This consisted of a 10" casing pipe with a 4" air line lowered down inside the casing pipe. The casing was securely anchored to the shaft timbers and could be easily extended as the water receded. Air pressure through the 4" pipe forced the water up through the casing pipe. Dewatering the shaft began on September 24, 1919. The water level could be held at a safe distance below the necessary retimbering operation.

Work slowed when the 400-foot level was reached, for it was necessary to recapture over 50 feet of the caved east drift. This was to make room for a 6,000-gallon pump storage tank and two Cameron turbine pumps. These pumps raised water directly to the drain tunnel. They were connected to 125 hp 440-volt motors, and were capable of lifting 500 gallons of water per minute. The first airlift, before it was abandoned, was raising the water 485 feet vertically, with submergence of 23 percent and a capacity of 250 gpm, with an air pressure of 67 pounds per square inch.

The airlift was placed in use again below the 400-foot level, and on December 31, 1919, the water was 550 feet below the collar of the shaft, and on February 29, 1920, it was 1,069 feet below the collar—or 47 feet above the 1000-foot level. The airlift had reached its limit, after which air pumps were used, and in two weeks the rest of the shaft and the 1000-foot level were pumped free of water.

Considerable timber had to be replaced during the shaft dewatering. Even though this was done, much of the other timber had deteriorated to a point where the shaft was rendered unsafe. The shaft from the 200-foot level to the 800-foot level had been run through stoped-out ground without having the protection of side pillars. The stopes had been backfilled with waste rock, and this fill had been held back from the shaft by stulls and lagging.

The shaft was originally divided into four compartments, but because of the changeover to self-dumping skips for hoisting ore and waste rock, the compartments needed to be as large as possible. The decision was then made to retimber the entire shaft, and at the same time to have just three compartments, each measuring 4 feet 5 inches by 5 feet in the clear. Port Orford cedar was used for all shaft sets, and spruce was used for the stulls and lagging along the shaft sides. Sets of 40-pound T rails were installed in each hoisting compartment to accommodate the skips.

The vertical pump shaft was the original shaft from which the mine had been opened, and it was located 150 feet west of the main shaft. The Cornish pump system was formerly used in this shaft to pump water from the 800-foot level to the drain tunnel. This shaft was in a dangerous condition, so a substantial bulkhead was installed above the 400-foot level. No effort was planned to hold the shaft open above that point. Much timber had to be replaced below the 400-foot level.

Reopening the Eureka Mine—1920

The Eureka mine had been worked from the mid-1850s until it closed in 1877. The grade of ore was very good, but during that early period the cost to mine and mill the ore ran between $12 to $15 per ton. Several miners who had worked there claimed that a considerable amount of $10 rock was used as fill material in worked-out stopes. They also claimed that rock of the same grade had been left in place. Based on this information, it seemed advisable for the company to reopen the Eureka mine. A good profit could be made on this grade of rock, which would help defray future development costs.

The water was removed from the Eureka mine simultaneously with rebuilding the Idaho Maryland surface plant and dewatering the Idaho shaft. There were 104 men working on the three projects. Water was pumped from both Eureka shafts at the same time, and the water level in the incline shaft was held at 35 feet below the 700-foot level. The timber in the incline shaft was in good condition. The timber had to be replaced at the collar of the vertical shaft, along with the ladders in the shaft.

Undergound connections were then made between the two mines from the Idaho Maryland 400- and 700-foot levels. Each level was extended out beneath the Eureka workings. A raise was put up from the Idaho 400-foot level connecting with the Eureka 300-foot level, then another raise was run from that level to the bottom of the east Eureka shaft. A raise was also run up from the Idaho 700-foot level to the Eureka 600-foot level, which was the lowest working level in the mine. These two connections provided mining access, second exits, and increased ventilation for both mines.

Systematic reclaiming of all Eureka levels took place, along with sampling the fill rock in all locations as they became available. Extensive stoping had previously taken place throughout the entire ore body, and the stopes had been backfilled, but the value of the fill rock was very low. Various sections of the mine that showed any promise at all were turned over to tributers to mine.

A persistent effort was made by the tributers to find enough mill rock to warrant continued mining, but to no avail. Having spent the time, effort and cost of reopening the Eureka mine, and then to find no ore, was disappointing. In fact, it was the most disappointing feature of the development program during 1920 and 1921. This venture proved two things: 1) that the sources who believed nearly 50,000 tons of $10-per-ton fill rock and unmined ore remained in the Eureka mine were unreliable, and 2) superintendent William Watt had not left any gold in his mine, over four decades earlier.

The Eureka incline shaft contained four compartments from the surface to the 700-foot level, and from there to the 1200-foot level the shaft was reduced to only two compartments. In cleaning out the shaft and pumping down to below the 700-foot level, it was found that a considerable amount of waste rock had been dumped into the lower section of the shaft.

Exploration of the Eureka ground in depth, below the Eureka shaft, could be accomplished from the Idaho Maryland, therefore further dewatering was discontinued. All operations regarding the Eureka were ended in the fall of 1921, and the incline shaft was allowed to fill with water up to the 600-foot level. Water above that level flowed to the Idaho shaft from that time on.

Idaho Maryland Surface Plant—1920

The surface plant, designed for 100 tons of ore per day, was nearly complete by the end of 1920. A crusher building, consisting of a 75-ton capacity ore bin and two 10" by 16" jaw crushers, was situated approximately 200 feet from the headframe. A 33-foot-high trestle was built to tram the ore from the headframe ore bin to the crusher building. A 16"

Two views of the Idaho Maryland shaft and headframe, along with overhead ore and waste trestles, as they looked in the 1920s.

belt conveyor was set in place to transport the crushed rock to the mill bins, a distance of 100 feet. The overhead trestle also extended from the headframe waste bin to the waste dump, for disposal of waste rock.

The blacksmith shop was rebuilt, and in addition to the original tools, a power hacksaw and a pipe-cutting and threading machine were added. The office building was shingled on all sides and roof, the inside lined with "beaverboard," and new floors laid. Two new shower baths, washbasins, and a hot-water system were added. Two new concrete vaults were constructed in the office.

A tailings-impounding dam was constructed on 7.57 acres of property purchased from John H. Hansen. There were 5,823 cubic yards of waste rock used in the construction. When completed, 1,400,000 cubic yards of mill tailings could be impounded. The plans for the tailings pond were approved by the California Debris Commission before construction began.

The 20-stamp mill was put in operation in November 1920, and up to the end of the year 555 tons of ore were crushed. The value per ton was $6.65, with the tails at 82 cents. The concentrates were accumulated for shipment to a smelter. The flow sheet was as follows: amalgamation on riffled iron chuck blocks in the mortars, through 35-mesh screens, over amalgamating plates, through two cone classifiers (with the overflow going to waste), sands to Deister sand tables, and the middlings returned and reconcentrated.

Mine Workers Protective League—1921

As previously noted, in June 1919 the miners of the district had called a strike against all the gold mines. There was a definite lack of leadership displayed by the striking miners. The Mine Workers Protective League was organized at that time to represent the workers and negotiate wage increases with the mine operators. A one-year contract agreement was signed at that time between the League and each of the mine companies. Neither side offered any changes to be made at the close of the first year.

In early June 1921 negotiations between the mine operators and the Mine Workers Protective League were opened. This meeting was to adjust wages and draw up a new agreement in place of the one to expire on June 30. Satisfactory solutions were not reached, so on July 1, 1921, the men did not report to work, except the hoistmen and pumpmen, who were allowed to work until July 5. Prior to calling out the pumpmen, the company had made arrangements for the United Comstock Co. at Virginia City, Nevada, to supply men to operate the pumps. The Comstock men showed up as planned, and there was no interruption in pumping operations.

Negotiations continued with the League until finally an agreement to arbitrate was drawn up and signed by both parties. The agreement stipulated that three arbitrators were to be chosen—one by the League, one by the operators and a third by the first two. In the meantime the miners agreed to return to work for $4.25 per day for miners and $3.75 for muckers and trammers, pending a decision by the arbitration board.

However, when the League's arbitrator and the operators' arbitrator could not agree on a third member, the agreement died a natural death. The League then approached the mine owners with another proposal, and an agreement was finally signed at the wages stated above. The contract period would extend to June 30, 1922.

The agreement also stipulated that time-and-a-half was to be paid for Sunday work, but it was understood that the Idaho Maryland Mine would be exempted from the extra pay for Sunday work until such time as the mine went into production.

The butcher shop that had been established as part of the 1919 labor agreement was discontinued in October 1921. The shop had been partially subsidized by the mining companies to provide meat to the miners at a reduced price. The money was well spent at the time, because it kept meat prices down and helped stave off more serious labor complications.

Underground Operations—1921

A considerable amount of dead work was necessary when the 1000-foot level was reached. First, a large sump, measuring 15 feet by 140 feet by 7 feet, had to be excavated from below the level. This sump provided a capacity of 110,250 gallons of water, and was capable of an 18-hour storage from all levels above and below the 1000-foot level, as well as all future water from the Eureka mine. Next, a pump station, measuring 20 feet by 25 feet by 12 feet, was excavated, and a 1,000-foot head, 250 hp motor-driven Taylor pump was installed. This pump delivered water to the surface drain tunnel through the previously used 12" Cornish pump column.

Previously, ore and waste had been hoisted in single cars on a cage. The changeover to self-dumping skips made it necessary to provide ore and waste pockets on all levels. Two parallel raises were put up in the hanging wall from the shaft below the 1000-foot level and connected with a crosscut in the form of a loop, to form an ore and waste pocket. A grizzly consisting of inverted 40-pound rail, spaced eight inches apart, was installed above each pocket to allow the rock to flow through the grizzly. Rock greater than eight inches would then be manually broken into smaller pieces by a 12-pound rock hammer.

A steel-armored 4,000 volt, 3-conductor cable, with a capacity for 600 hp was installed in the Idaho shaft down to a transformer station on the 1000-foot level. Three 150 kva transformers, with a capacity to operate the underground pumps and hoists, were set in place on the level.

Preparations were underway to dewater the lower levels concurrently with all the other work taking place on the 1000-foot level. The Canyon shaft collar had to be retimbered, and the station enlarged for future ore and waste bins. A temporary air hoist was then installed on the station to begin dewatering operations. The airlift method was again used, and proved satisfactory. Heavy ground was encountered for 125 feet below the 1000-foot level and required a lot of new timber. The shaft was in fairly good condition below that point. Electric pumps were installed on the 1300-, 1600- and 2000-foot levels as the water receded. The old Cornish pump column, which was still in place, was used in conjunction with the new pumps to raise the water to the 1000-foot level sump. The Canyon shaft was free of water to the 2000-foot level on June 29, 1921. It wasn't until October 9, 1921, that the 1900-foot level in the Dorsey winze was clear of water.

A 75 hp double-drum hoist was placed on the 1000-foot level, 100 feet in the footwall, in line with the collar of the Canyon shaft. Raises were run to provide clearance for the hoist cables and sheave wheels. A double compartment bin was erected at the collar of the shaft and a double track of 20-pound T rails was laid from top to bottom in each hoisting compartment, a distance of 2,059 feet.

Surface Operations—1921

On January 27, 1921, a fire broke out in the dry building, and before the water could be turned on, the whole lower floor was ablaze. The fire was put out by using two high-pressure monitors, one in front of the building and one at the side. Similar monitors had previously been placed for protection of all buildings. The miners had to wear their mine clothes home, because their street clothes were destroyed.

A 40-foot derrick mast and a 50-foot boom were erected near the shaft collar for loading materials in and out of the skips. The derrick was equipped with a small compressed-air hoist for raising or lowering material. The derrick could swing in a large arc, thus allowing items to be picked up as much as forty feet from the shaft collar.

Poles, when delivered, were offloaded at the top of the waste dump and allowed to roll down the slope in easy reach of a large swing saw, sawmill, and the large derrick hoist. The sawmill was capable of cutting lumber for making mine ladders, lagging, and heavier timber used for shaft and raise sets. These could be cut much cheaper than by off-plant purchase.

The only change made in the mill was to cut out classification and send all the pulp over the Deister sand tables, and the middlings over a Wilfly table. This change materially lowered the tailings loss, but plans were to add another Wilfly table so that the middlings could be classified, the coarse going to one Wilfly table and the fines to the other. This move was planned to reduce the amount of loss in the tails.

The Grass Valley community succeeded in establishing a very fine recreation park as a memorial to those young men who served and were killed during World War I. The mine companies were approached to see what they could do toward making the park a reality. The Empire mine donated the land, bounded by Colfax Avenue and Race Street. The Idaho Maryland Mines Co. built the community building situated in the middle of Memorial Park. The cost for building the structure was $2,500.

Underground Mining—1922

The upper levels above the 1000-foot level were not considered to be favorable for finding ore. Several tributers wanted to try, and they were given that chance. This part of the mine was thoroughly stoped; originally by the Coleman brothers between 1865 and 1893, and again when tributers worked the same area extensively during the period between 1904 and 1914 under the Idaho Maryland Development Co.

One area where it was possible that ore remained was a block of ground below the old pump shaft that terminated on the 800-foot level. A two-compartment raise was run from the 1000-foot level to the sump of the old pump shaft. The raise did not find any ore, however it provided other advantages. Water from the Eureka mine was able to be piped down the raise directly to the 1000-foot level sump, rather than

The community building that the Idaho Maryland Mines Co. built in the middle of Grass Valley's Memorial Park at a cost of $2,500.

having to pipe the water down the shaft. This raise also provided a second exit directly to the surface, through the pump shaft.

The east drift on the 1000-foot level was cleaned out, then driven farther eastward to ascertain if any parallel veins existed, or whether the Dorsey vein extended upward to this horizon. At a point 1,375 feet from the Idaho shaft the quartz stopped. Even though the quartz pinched out, a strong fissure continued. The drift was driven further until it was stopped a distance of 2,515 feet from the shaft. No other quartz veins were discovered.

At the recommendation of Wilbur H. Grant, a University of California geologist, a crosscut was driven due south from the 1000-foot level east drift to get under the Black Hawk country and intersect veins showing on the surface. The crosscut was started at transit point 202, approximately 200 feet back from the breast of the east drift. Between 50 and 1,015 feet the crosscut was entirely in massive porphyrite. Between 1,015 and 1,710 feet there was a distinct schistose development of amphibole schist. Between 1,710 and 2,045 feet (the breast), the massive porphyrite reoccurs. There were no vein structures of any significance found.

A mule barn was constructed on the 1000-foot level and two mules were used for tramming cars, due to the long distances involved in the east drift and south crosscut. This was the only level in which mules were used in the Idaho Maryland mine, and then for less than two years.

The 1300-foot level east drift was driven 210 feet before being stopped, due to discouraging results. The 1600-foot level was reclaimed from the Canyon shaft to the Dorsey winze, a total distance of 1,995 feet. Many of the workings in the Dorsey winze area were open, as well as the 1500-foot level above. A good-looking vein was found on the 1900-foot level of the Dorsey winze. The drift was about 60 feet long, with the quartz averaging 15" wide and assaying $20 per ton. The vein measured 38" in the face and assayed $36. This was the only place in the mine where hearsay evidence about ore conditions was verified.

The existing drifts on the 2000-foot level from the Canyon shaft were made ready for further exploration. By the end of 1921 the west drift was cleaned out and repaired to the face, a total of 303 feet. A strong fissure was found to exist, and in places the vein formation opened up to eight feet in width, but on the whole the vein was very erratic and very low grade. The gangue minerals consisted entirely of calcite, dolomite and allied minerals, but no quartz.

At 189 feet a fault dipping steeply southwest threw the vein 8 to 10 feet, and at this point some good values were found. A seam of goudge on the fault plane assayed about $100 per ton. It was believed that this was the same fault plane that was exposed in the 1800-foot level west drift, 50 feet from the Canyon shaft, and along which a rich ore shoot was mined by the old Idaho Mining Co.

By the end of the year the east drift on the 2000-foot level had been driven 198 feet, for a total distance of 231 feet from the Canyon shaft. A very good showing of quartz was exposed, which, although very erratic as to size and value, gave strong evidence of ore possibilities below the level.

Water from the lower mine levels was raised to the 1000-foot level sump in the following manner: water from the Dorsey winze sump was pumped to the 1800-foot level sump, and from there to the 1600-foot level by means of two separate small air-powered sinking pumps. The water from the 1600-foot level was piped down the Canyon shaft and was used as an ejector to raise the water from the bottom of the shaft into a 25,000 gallon capacity sump on the 2000-foot level. A 125 hp, 300 gpm, 620 foot-head Cameron turbine pump raised the water from the 2000-foot level to the 1500-foot level sump. From there the water was pumped to the 1000-foot level sump by a 125 hp, 300 gpm, 585 foot-head, Cameron turbine pump.

Additional Properties Acquired—1922

During the development period the Idaho Maryland Mines Co. had been exempted from paying time-and-a-half for Sunday work by the Mine Workers Protective League. The exemption was lifted, beginning on August 1, 1922, because the mine was producing ore. Concurrently with the new requirement, the company discontinued all Sunday work except routine duties performed by hoistmen, millmen and pumpmen.

Subsequent to the Metals Exploration Co. taking control of the consolidated group of mines in 1919, several other properties were acquired. These consisted of the Grant claim, East Eureka claim, Bastian homestead, Vincent homestead, and the John H. Hansen property. The company purchased controlling interest in the Black Hawk mine, and obtained patents covering the East Eureka and 15 other claims.

An agreement was signed with the Brunswick Consolidated Gold Mining Co. for a right-of-way underground from their property line westerly to within their extra-lateral rights line under the Union Hill group of claims. This was granted in exchange for a quiet title to two fractions to which both companies had claimed ownership. The exchange of the right-of-way for the two fractions was considered a great bargain.

Underground Operations—1922

The development work undertaken during the year was widespread. Three short crosscuts were driven in the footwall of the 1000-foot level east drift, none of which revealed any parallel veins.

Three short crosscuts were driven in the footwall of the 1600-foot level east drift. Only one showed any encouragement. Crosscut No. 760 was 83 feet in length, all of which was in serpentine. A well-defined fissure, lying parallel to the main vein and dipping 72° to 83° to the south, was encountered about 35 feet from the east drift. Short drifts were driven both east and west, the fissure exposing only intermittent quartz, ranging up to eight inches in width. Two short crosscuts run in the hanging wall revealed only the diabase dike.

The old drift east of the Dorsey winze on the 1600-foot level was reopened to a point 290 feet east of the winze, exposing an old stoped area throughout. At a point 15 feet back from the breast, the vein entered a diabase dike that crossed the drift diagonally. The quartz dissipated rapidly in a fan of stringers in the dike. A crosscut started at this point was run 188 feet in the footwall, where a drift was driven 196 feet. Most of this drift consisted of low-grade quartz stringers in diabase. The 1600-foot level was quite disappointing for finding ore.

The Dorsey vein on the 1900-foot level was driven eastward, following strong quartz with good values for 130 feet. At that point the quartz was dragged to the left by a series of faults. The vein disappeared and many flat quartz stringers branched off into the footwall. The main vein was picked up again at a distance of 30 feet by crosscutting due north. The vein was then drifted on easterly, exposing 35 inches of quartz that assayed $10.80 per ton. The quartz pinched out at 130 feet, but the strong fissure continued. At a point 175 feet in, a crosscut was driven northerly, and another quartz vein was encountered. This vein was followed for 125 feet, exposing good quartz throughout. At the east breast, the quartz cuts through a diabase dike at a dip of 50°, whereas the dike had a dip of 68°, the strikes being parallel.

At the westerly breast a stope was started where the quartz measured five feet in width. This was in hard serpentinized gabbro, and was considered to be one of the strongest showings in the mine since it was reopened. Ore taken from the Dorsey winze area and sent to the mill

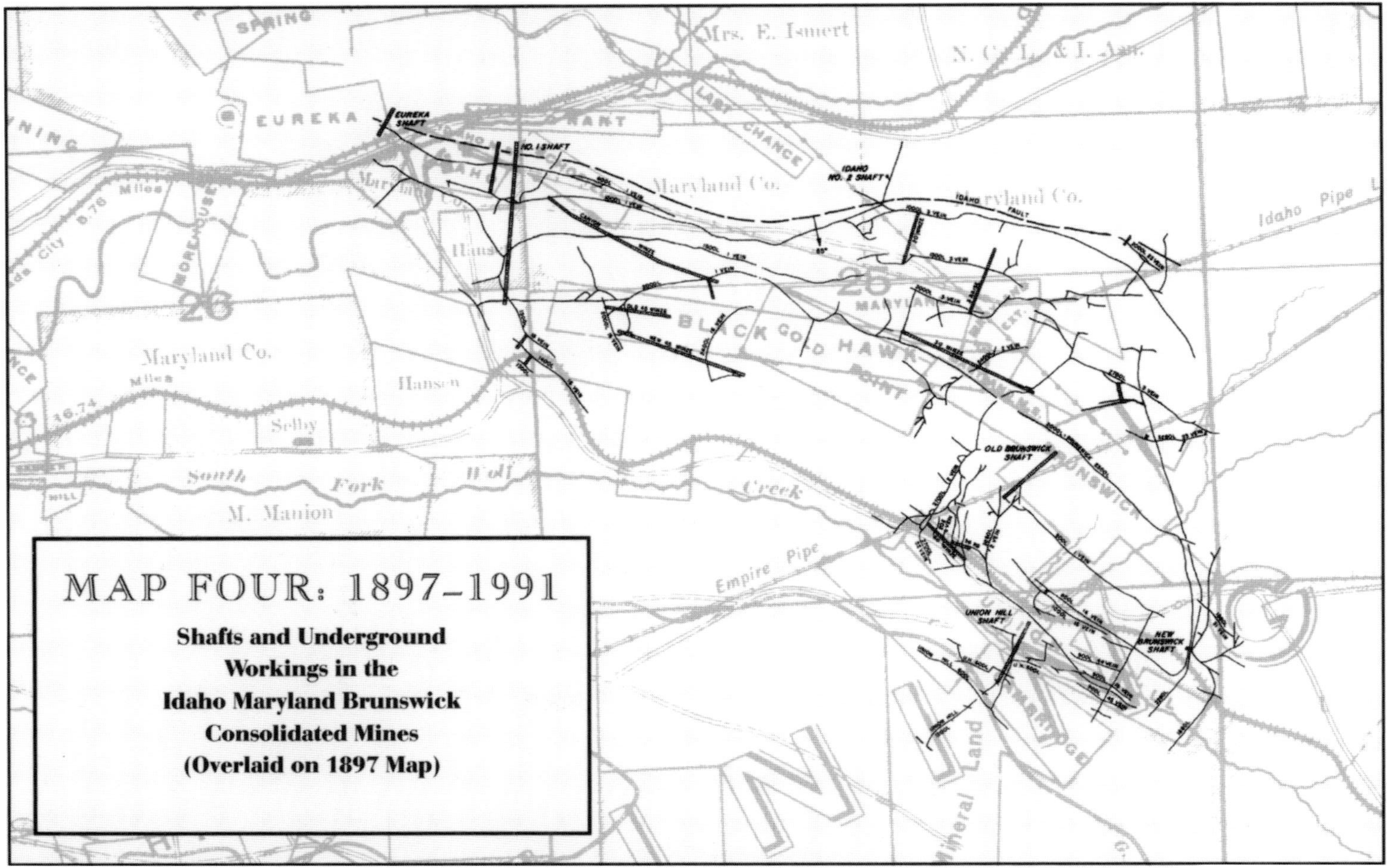

MAP FOUR: 1897–1991

Shafts and Underground Workings in the Idaho Maryland Brunswick Consolidated Mines (Overlaid on 1897 Map)

during 1922 totaled 4,271 tons. Plans were to sink the Dorsey winze to the 2,000-foot level elevation, however, after sinking only 50 feet the sinking operation was stopped.

The Canyon shaft was sunk an additional 63 feet, the bottom of which measured 2,122 feet on the incline below the collar. The 2100-foot level was then opened up and drifted on for a total of 654 feet. A strong but narrow vein followed the diabase-serpentinized gabbro contact. The quartz was less than 12 inches wide for most of that distance. Two short crosscuts were driven from the west drift, one in the footwall and one in the hanging wall. Nothing was discovered in either one. The No. 87 raise was run up from the 2100-foot level and opened into the 2000-foot level at the Canyon shaft station. There were 2,664 tons of ore sent to the mill from the 2100-foot level.

The 2000-foot level west drift was advanced 1,338 feet, for a total distance of 1,742 feet from the Canyon shaft. This drift did not expose sufficient quartz to warrant any development for ore. The 2000-foot level east drift was advanced 1,523 feet, for a total distance of 1,754 feet from the Canyon shaft. Very little quartz occurred, the greatest width being four inches.

The Canyon shaft station on the 2000-foot level was enlarged and a hoist set in place. The No. 87 raise was then utilized for sinking a winze, beginning at the bottom of the raise on the 2100-foot level.

In early 1922 work commenced to deepen the Idaho shaft. The main purpose of extending the shaft was to explore the Eureka-Idaho-Maryland vein below the 1000-foot level, in hopes of finding another ore shoot. The eventual connection with the 2000-foot level would eliminate triple handling of ore through the Canyon shaft, and facilitate the movement of ore, men and supplies direct from the 2000-foot level. This shaft would also provide a second exit from the lower levels, as well as provide increased natural air circulation.

Between 80 and 95 feet below the 1000-foot level the main hanging-wall dike developed a reverse dip to the north, across the shaft. The shaft then continued in serpentine, quite schistose near the dike, but becoming more massive in depth. From 200 feet to a point 655 feet below the 1000-foot level a light-colored dioritic rock prevailed, interbanded with dark serpentine or serpentine schist in variable quantities. This formed a comparatively hard ground from which to sink.

The diabase dike returned to a southerly dip and crossed the shaft again between 655 and 740 feet below the 1000-foot level. The strike averaged north 66° east, and dipped 30° to the southwest. Quartz developed in the footwall of the dike, varying from nothing to 18 inches, but averaging about 10 inches. This was accompanied by considerable soft serpentinized gouge.

Surface Operations—1923

The surface plant was in good operating condition, much of which was relatively new. The sawmill was very active in preparing timber for all underground needs. Spruce and pine were sawed, the pine cut from the mine property, and the spruce purchased from an outside contractor.

On June 30, 1923, a new one-year agreement with the Mine Workers Protective League went into effect. This carried a 50 cents per day increase for all employees. The pay increase benefited those on day's pay more than many of the underground workers. Those on development work were on a contract basis, and these miners made better than day's pay, as they were paid for the footage broken. At this time the company placed much of the stoping on a contract basis also, in hopes of increasing the tonnage of ore. This did not help much, for there weren't enough stopes available to make much of a difference. There were 140 men working at the mine at that time.

The mill operated only part of the time, and then sometimes just 15 stamps were used. The intermittent ore production resulted in higher milling costs. Many suggested changes to be made in the mill had to be put off until such time as the mine was producing a sufficient amount of ore to warrant these expenditures.

Underground Operations—1923

During the years 1923 and 1924 the underground operations involved work in the lower levels of the mine. Midway between the 1000- and 2000-foot levels, the 1500-foot level was cut. A shaft station with ore and waste pockets was prepared. Drifts were started both east and west from the station.

A vein that was discovered while sinking the shaft looked encouraging, particularly as it appeared to bring within the range of possibility the country underneath the Eureka shaft. There was 432 feet of drifting on this vein, but absolutely nothing in the way of values was found, so the work was stopped.

The vein was the so-called Morehouse vein, but as a matter of fact, it was nothing more than the Eureka vein on an approximate north and south course, instead of an approximate east and west course. The change of strike was due to a diorite mass which changed the course of the fissure. The continuity of the vein was unbroken, although it changed its strike from north 50° east to south 20° east in less than 400 feet. Thus a trough was formed which plunged into the earth with a dip of 30° to 35° from the horizontal, and a strike about 60° east. This strike was roughly parallel to the axis of the main ore shoot, but with a dip of about 10° steeper than the angle the rake of the ore shoot makes with the horizontal.

Sinking the No. 87 winze began on the 2100-foot level in line with the No. 87 raise, which was utilized as the upper portion of the winze. It extended downward for 229 feet below the 2100-foot level station, where ore and waste pockets were completed on the 2350-foot level. This level was developed by the east drift 612 feet beyond the winze, and the west drift driven 650 feet. The drifts were driven on the diabase-serpentine contact, and the quartz vein exposed was low grade and could not be mined at a profit. A crosscut was driven 125 feet into the footwall of the west drift, but did not reveal any vein structures. The No. 7 raise was run up 149 feet from the west drift.

The 2000-foot level west drift cut the No. 935 vein 950 feet west of the Canyon shaft. Drifts were then driven 48 feet in a northerly direction, and 375 feet in a southerly direction. This vein was believed to be the Morehouse vein, as it was found underlying the same diorite mass and had the same general dip and characteristics as the vein explored on the 1500-foot level. The values and vein widths were quite erratic, but on the whole were encouraging enough to warrant further exploration.

The No. 45 winze, located 50 feet south of the main west drift, was sunk 538 feet on this new vein to explore its downward dip. The winze connected with the top of the No. 7 raise that was previously put up from the 2350-foot level. Stations were cut on the 2300- and 2500-foot levels, and short drifts driven on each side of the 2300-foot level station. A total of 2,660 tons of ore from this vein was sent to the mill with an average yield of $10.90 per ton.

In early 1923 the No. 1923 crosscut was driven in the footwall of the 2000-foot level east drift, where it exposed the No. 80 vein. The vein was explored above and below the level. It was thought to be the downward extension of the old Idaho Maryland ore shoot. Plans were to drive a raise through to the 1700-foot level, and then continue on to connect with the bottom of the 1600-foot level prospect winze. The raise was driven diagonally up the vein in order to follow the supposed general trend of the ore shoot.

However, the quartz vein disintegrated 100 feet below the 1700-foot level elevation, and lost its identity in the diabase-serpentine formation. Several branch raises were driven directly up dip from the No. 80 raise, each extending 60 to 80 feet in length, before the vein pinched out. A fairly large tonnage of $10 rock was removed.

Soon after the No. 1923 crosscut cut the No. 80 vein, it also cut the No. 465 vein and, as anticipated, it proved to be the Dorsey footwall vein. It did not carry the values that existed on the level above, nor did it have the same characteristics. The mineralization consisted entirely of stringers in diabase, whereas the vein above was a massive quartz vein, from two to six feet wide. The drift was driven 144 feet west of the crosscut and 644 feet east of the crosscut.

The only values found in this drift were from 500 feet to 540 feet east of the crosscut in diabase stringers averaging $10 per ton. The drift when stopped had less than 300 feet to go to reach the western boundary of the Independence quartz mining claim. A crosscut was driven in the hanging wall 264 feet, and a branch vein was drifted upon for 132 feet, revealing no profitable ore.

The No. 89 raise was run up to where the 1950-foot intermediate level was established, and a drift 231 feet in length was driven. The No. 89 raise continued and connected with the 1900-foot level.

In the latter part of 1923 a prospect drill was purchased and placed in immediate use. This core drill was a great innovation in mining, and gave the same geological results with a small diameter core as running a 5-foot by 7-foot crosscut. This type of exploration proved very successful, with a great savings in time and money. Eighteen separate holes were drilled, ranging from 31 to 206 feet in length, for a total of 1,339 feet. The areas prospected were on the 2000-foot level No. 465 drift, 1500-foot level near the Idaho shaft, and on the 2350-foot level.

Management Change—1924

On February 12, 1924, a change in management was effected. M. A. Roche became mine superintendent, replacing John A. Fulton, and on May 10 a change in mine foreman and other underground bosses took

John Fulton, manager of the Idaho Maryland mine, and John Glasson stand behind "Aunt Helene" and Bernice Glasson as they prepare to descend into the mine sometime in the early 1920s.

place. Much better conditions were obtained after stable relations were established. Superintendent M. A. Roche wrote in his annual report:

> "The efficiency of labor was very low, particularly during the summer months and toward the latter part of the year. The men underground have a fast and set rule as to what embodies a day's work. With this set rule of timed effort on the men's part and the well known evil of excessive high-grading at every opportunity, a constant vigil must be kept at every working place. Because of this, an excessive number of bosses must be employed. The above conditions result in high operating costs and low labor efficiency."

Several miners were riding up a winze on a manskip when it was suddenly derailed by loose rock on the track. Thomas Reed was thrown off and rolled down the incline a distance of 50 feet, sustaining serious injuries. None of the other men was injured. The accident occurred at 3:00 p.m. on April 23, 1924.

Reed was brought to the surface as soon as possible. Dr. Carl P. Jones had been summoned, and upon arrival it was discovered that the injured man had received a broken neck with complete paralysis and severe bruises to one side of his body. Thomas Reed died that evening in the Jones Memorial Hospital.

Sinking in the main shaft continued on schedule. The shaft was sunk in serpentine schist formation from the 1500-foot level to the 2000-foot level, which required very little timber other than to carry the track along the footwall.

The Idaho shaft completion time was shortened by work accomplished on the 2000-foot level. A large shaft station was cut out adjacent to the west drift, and a large area was widened for grizzlies. At a point where the shaft would intersect the 2000-foot level station, a winze was sunk to a depth sufficient to allow for ore and waste pockets and a sump below. A raise having the same dimensions as the shaft was run upward, in line with the shaft, for a distance of 230 feet.

The shaft sinking was completed in June 1924, and ore and waste was then hoisted directly to the surface from the 2000-foot level. Prior to this time the ore and waste from below the 1000-foot level had been hoisted through the Canyon shaft, with the added expense of double and triple handling. The Canyon shaft continued to be used for pumping purposes.

The main hoist was found to be inadequate for hoisting rock directly from the 2000-foot level. Larger drums, reduction gears and larger bed plates were added. The 150 hp motor was replaced by a 300 hp motor, thus doubling the capacity. New one-inch-diameter cables were installed on each drum. Many improvements were made to the headframe, such as tie rods, additional timbers and supports. The storage capacity in the headframe bins was also increased.

Good ore continued to be mined on the 1900-foot level of the Dorsey winze during 1924. Two of the quartz veins were mined from the 1900-foot level up dip for over 150 feet. A fault was encountered at that elevation, and the vein pinched out.

All operations in the Dorsey winze area were stopped in June, when hoisting through the Canyon shaft was discontinued. Plans then were to work the Dorsey winze ore through future raises from the 2000-foot level.

The footwall veins on the 2000-foot level were explored by the No. 465 east drift, the No. 94 west drift, the No. 95 west drift and the No. 200 crosscut. Several long core-drill holes were also drilled in the serpentine footwall, which were to pick up any new quartz veins.

A crosscut was driven 377 feet in the hanging wall on the 2000-foot level, 1100 feet east of the Canyon shaft. Nothing of any value was discovered.

The No. 80 vein was explored in its downward course below the 2000-foot level by the No. 30 winze. The vein appeared to pitch directly into the Brunswick mine property, so it was necessary to obtain an option from the Brunswick Consolidated Gold Mining Co. The option was granted on September 9, 1924, at no cost other than the payment of the county taxes on the land involved.

The two-compartment winze was sunk diagonally down dip of the vein for 900 feet. Five levels were established, and drifts driven on the 125-, 260- and 375-foot levels. Stoping occurred on each of these levels. The best ore came from the 260-foot level, which averaged $10.90 per ton. The vein pinched out at 805 feet below the collar of the winze. The 625- and 850-foot levels were not developed, the vein being only a seam.

A total of 10,192 tons of ore was mined from the No. 80 vein, with an average of $8.15 per ton. Work in the No. 30 winze was disappointing, because of the low grade of rock.

During 1924 there was 7,365 feet of development work done in an effort to discover new ore bodies. The cost of development was $249,054. The gold recovered from stoping had a value of $53,709. The loss for the year was $260,906.

Mine Closing—1925

On February 7, 1925, all work in the No. 30 winze area was stopped and the Brunswick Consolidated Gold Mining Co. notified to discontinue the option.

Superintendent Roche was disappointed that development throughout the mine did not discover any new large ore bodies. He felt that those found contained ore in such small quantities that they could not be the basis of a profitable mining operation. He also felt that further expenditures upon the property were not justified. Upon his recommendation, the mine was shut down in April 1925, and was to fill with water.

During the six years of operation under the Bulkely Wells interests, over 10,000 feet of caved drifts were reclaimed, three prospect winzes sunk, 25,000 feet of drifts and crosscuts driven, and the Idaho shaft extended 1,900 feet to the 2000-foot level.

In 1919 the control of this company had been acquired by Wells for the Metals Exploration Co. From January 1919 to September 1925 over $2,250,000 was expended on the purchase, reopening and development of the consolidated properties. During this same period only $200,000 was taken out in the course of operations.

Errol MacBoyle.

CHAPTER 8: IDAHO MARYLAND CONSOLIDATED MINES, INC.

Following the Idaho Maryland mine closure in 1925, Errol MacBoyle set out to raise money to keep the pumps running, to prevent the underground workings from filling with water. His faith in the mine was steadfast, and his perseverance was equally as strong. MacBoyle felt certain that the mine could once again become a major gold producer if operated properly. He had definite theories in regard to the vein and fault systems, which the previous operators chose to disregard.

Edwin Oliver and Frederick McNear, longtime associates of MacBoyle, were the only persons willing to advance money to keep the mine operating, and for every dollar they invested, MacBoyle offered them one share of preferred stock, plus two shares of common stock as a bonus.

First a holding company was formed to acquire the holdings of the Metals Exploration Co., which consisted of over 700 acres of land and included the Eureka, Idaho Maryland, Union Hill, Gold Point, Black Hawk and South Idaho mines. The purchase price was in excess of $2,500,000.

The holding company was incorporated in the state of Nevada for three million dollars and was called the Idaho Maryland Consolidated Mines, Inc. The company's officers included Edwin Letts Oliver, president; Frederick W. McNear, vice president, Errol MacBoyle, managing director; and directors L. H. Duchak, Paul Claiborne and Judge Edgar T. Zook.

The operating company, charged with running the mines, was headed by MacBoyle as president and general manager, and its directors included MacBoyle, Oliver, McNear, Rufus Thayer and Roy H. Elliot.

For six months prior to the cessation of work in April 1925, the available rich ore had been "gutted out" in anticipation of the closure. Good ore bodies were subjected to poor mining methods, drifts were allowed to cave, and equipment allowed to deteriorate. It was the general opinion of E. M. Roche, the underground superintendent, that the mine was worked out.

Before notification was given to the minority directors of the intention to close the mine, the No. 30 winze, No. 45 winze and No. 87 winze were allowed to fill with water. Most of the equipment from the 2000-foot level had been removed, and timber repair neglected in several areas. Further deterioration of underground workings took place between the time operations were suspended and when the new Idaho Maryland Consolidated Mines company took control.

Start of Operations—1926

When actual operations began on January 1, 1926, under the management of Errol MacBoyle, it was estimated that it would take six months and $50,000 to open the underground workings so that ore bodies would become accessible. It also was estimated that an additional six months and $50,000 would be needed as an operating fund to carry the properties until enough ore could be produced to keep the mill working at its capacity of 2,000 tons per month.

Fortunately, most of the miners who had worked in the mine were anxious to secure "tribute pitches" or leases. This made it possible to start operations under a tribute system, which expedited the reopening of the mines at a cost substantially less than the estimated figures. Twenty-five of the best workmen were selected from the men who applied for tribute pitches. These were divided into six leasing companies, and each group was given a specific block of ground on three of the six known veins. The other three veins were reserved for company operations.

An agreement was entered into whereby the 25 men, as part consideration for their leases, agreed to furnish all the labor to reopen the 2000-foot level. This included replacing damaged timber when necessary, and reinstalling the pipe, rail and trolley wire between the shaft station and the end of the level, a distance of over 4,000 feet. In addition, the tributers agreed to pay for half the powder used, and all of their worker's compensation insurance.

The work of reopening the 2000-foot level was started on January 20, and by the first of March the electric trains were running to the far eastern end of the level. This would have taken three months or more to accomplish on day's pay. By March 1 the tributers had started in their individual stopes, and within a month they began producing ore. At this point, the outlook for the future of the mine looked promising; the mine had been rejuvenated for the small sum of $18,000.

The Mike McGuire tribute was given a block of ground on the No. 80 vein on the 2000-foot level. This group was able to open the stope in record time and gradually increased production. The vein averaged 6 to 8 feet in width and from 2 to 4 feet of quartz showing heavy free gold at intervals. The stope was opened for a length of 125 feet along the strike of the vein, and the ore produced ran from $17.23 to $31 per ton in gold recovered by amalgamation alone.

The stope produced many sample sacks of specimen gold throughout the year, and the face was still showing heavy gold in a portion of the stope at the 1900-foot level elevation of the Dorsey workings. The McGuire group had reached the upper limits of their lease and were forced to start over at the bottom of the remaining portion of the stope.

William Yonker was assigned the 1950 vein, No. 89 vein and the east Dorsey vein on the 1900- and 1950-foot levels. These workings were in bad condition, and it took longer to begin producing ore. There was a lot of development work involved and it was hard to eliminate a mixture of waste rock from the ore. All three veins were 4 to 8 feet between walls with 2 to 4 feet of ribbon quartz, showing free gold and galena. This group was able to keep five stamps operating much of the time. The ore averaged $8.63 per ton.

The E. R. Stambaugh miners were assigned an area in No. 45 winze on the 2000-foot level. There was considerable dead work necessary, which involved dewatering the winze, retimbering some sections and clearing waste rock from the footwall timber as the water receded. The previous operators had dumped waste rock down the winze.

The 2230-foot level was cleaned out and the drift extended 100 feet on the Morehouse vein. No attempt was made to start stoping until more exploration was done. The first mill run of development ore averaged

$14 per ton by amalgamation only. The vein opened up a fine showing of coarse free gold.

The William Harvey group was given a block of ground on the 2000-foot level which produced only low-grade ore. In order to compensate this group for their labor in doing company deadwork, they were given a block of ground on the 1600-foot level No. 160 and No. 180 veins. The No. 180 vein ranged from 2 to 4 feet of good quartz showing heavy gold. A raise was put up 90 feet to explore the vein.

The Albert Sears miners were assigned a block of ground on the 2000-foot level on the No. 80 vein. The ore from this stope averaged $14 per ton with development waste mixed in.

The Cecil Norris group of miners was given a block of ground above the 2000-foot level on the Morehouse vein. These miners continued raising and stoping. The mixed ore and waste from this area ran about $7.80 per ton.

The tributing program proved satisfactory, and by June 1, 1926, there has been 2,004 tons of ore milled, with an average value of $13.45 per ton. This enabled the mill to start sooner than anticipated, and came close to placing the property on a self-supporting basis. Each of the tribute groups continued to produce well in terms of tonnage. They had to be reminded periodically to keep their development work current so as not to mine all of the good ore at once.

MacBoyle had to be away from the mine for long periods of time in pursuit of financing to fund the necessary development work required to find new ore bodies. A knowledgeable and capable person was needed to oversee and direct operations in his absence, so he hired Albert Crase as superintendent. Bert Crase had been with the Orleans Mining Co. for eighteen years, culminating with his appointment as superintendent of the Sultana mine.

The company work consisted of exploratory work on the 1000- and 1500-foot levels. A previously run crosscut located on the 1000-foot level, a short distance east of the shaft, was extended in the footwall.

The 1500-foot level east drift was driven from the shaft to where it connected with the 1500-foot level west drift previously driven from the Canyon shaft. The vein showed some free gold, but was too narrow for profitable mining.

Retimbering the Idaho shaft continued from October through December. The work was done during the night shift and graveyard shift,

Albert "Bert" Crase (left) at the collar of the Idaho Maryland incline shaft.

so as not to interfere with hoisting ore on day shift. Work was slow in some areas where it was necessary to remove some heavy slabs of rock left hanging when the shaft was retimbered in 1920.

From the time the mine began operations in March until the end of December 1926, there were 10,415 tons of ore milled. The gold recovered by amalgamation and concentration amounted to $118,030. The average value of the ore ran approximately $12 per ton, which was remarkably high, in view of the fact that at least 70 percent of the rock milled was low-grade development ore.

The big concern was the high loss of gold in the mill tailings. The samples ran $2 per ton when low-grade rock was being run, and when higher grade ore was milled the tailing samples ran from $2.50 to $6 per ton. The large tailings loss was primarily attributed to the fine slimes containing galena, chalcopyrite and other sulphides carrying high gold content, which was lost in existing concentration techniques.

Tributers Operations—1927

The tributers program proved a savior for the continuation of the mine during 1927. Most of the ore was mined by the tributers, which kept the mill running nearly all of the time. They had to include development rock along with the ore from stoping operations, and this caused considerably dilution of waste rock with the ore. Even then, the yield of gold was more than satisfactory.

The James Woods tribute was located on the 2225- and 2350-foot levels. These partners had to start at the 2000-foot level and pump the water out of the No. 87 winze to reach the two levels. Reopening the winze was slow, due to the need to replace a lot of timber. These two levels had to be reclaimed before mining could begin.

The vein was only 6 to 12 inches in width. The quartz contained heavy gold and galena, and when milled it averaged better than $12 per ton. The ore milled from these two levels contained about 75 percent waste rock. A lot of specimen rock was found in this narrow vein. James Woods and his partners had a disagreement, and the lease was turned over to Len Gillette.

The E. R. Stambaugh group in No. 45 winze area found fair mill rock on the 2230- and 2300-foot levels. There was very little quartz on the 2230-foot level, but free gold was present in the altered Morehouse vein material. Drifting on the 2230-foot level opened up a vein three feet wide, showing heavy specimen gold at times. The ore from this level ran from $10 to $28 per ton.

The Mike McGuire miners in the No. 80 stope on the 2000-foot level were the most successful in terms of production and also ore values. In addition to stoping, these partners ran two raises and drove two intermediate drifts within the stope itself. A raise driven in the footwall of the stope discovered a large body of rich ore directly in the footwall of an old raise run during the Fulton regime.

This development work proved the theory that the No. 80 vein was directly connected by a "linked" system of veins with the Dorsey and Dorsey footwall veins. The "H" intersection of these two veins was more or less masked by a broken zone and unless the vein structure was understood it would be easy to miss large bodies of ore. If the wrong fracture was followed it would give the impression that the vein and ore body was pinched out. The ore mined in this area was stoped from 6 to 8 feet wide, and although the quartz was only two feet wide, the altered diabase and serpentine carried coarse gold and heavy sulphides.

The Mike McGuire group of tributers had stoped beyond the area outlined in their agreement, and their time limit had expired. The lease was then taken over by the company account on April 1, 1927.

The William Yonker tribute continued development of the No. 1950 vein from the No. 97 raise between the 2000- and 1950-foot levels. The vein measured four to six feet between walls showing gold and galena, and was good milling ore above the fault. The vein was evidently lost by former operators, because their raise continued over the ore into the hanging wall of the vein. Yonker's group ran a raise on the No. 1950 vein over 150 feet, above the 1900-foot level.

They continued stoping on the 1900-foot level east Dorsey vein from the No. 89 raise between the 2000- and 1950-foot levels. The quartz in this area widened and ran well in the mill. This work proved that a number of large intersecting veins, which ranged from 6 to 9 feet in width, were in the footwall of the Dorsey winze. Everything between walls was sent to the mill, including the vein material. The Yonker tribute was taken over by the company in September 1927.

The William Harvey lease continued working the No. 180 vein until July. They sank a winze a short distance below the 1600-foot level east of the Canyon shaft, and also drove a raise above the level. Unfortunately, no ore was found in either location.

The Cecil Norris miners continued raising and stoping on the Morehouse vein above the 2000-foot level. The quartz ranged from 6" to 24" in width, with a quartz stringer zone 6 to 9 feet wide. The ore averaged about $8 per ton, with intermittent coarse gold in the vein, and gold showing in the stringers.

The William J. Mow tribute ran for only a few months on the 1000-foot level about 300 feet east of the Idaho shaft. This group worked on reopening an old stope on the eastern end of the original Idaho Maryland ore shoot. The rock milled included fill rock, pillar rock and a lot of serpentine filling. The mill recovery was better than $8 per ton. There was not enough rock to sustain continued operations.

Company Operations—1927

Timber crews were kept busy replacing timber sets in the Canyon shaft and on the 2000-foot level. This work was done on the night shift so as not to interfere with ore production.

The dry building burned to the ground during the night of March 7, causing the loss of the structure and the miners' clothes. The Grass Valley volunteer fire department responded, but the fire was beyond their control. Efforts were directed toward protecting adjacent structures. Construction of a new building began right away, and the miners were compensated for their clothing loss.

Over the period of four months several flotation engineers were called in to conduct a series of experiments to determine how to reduce the high loss in the mill tailings. The consensus of these men was to add additional Frue vanners to capture the finer sulphide particles from the slimes. Four new vanners were installed, resulting in a better recovery. These engineers were also of the opinion that the Idaho Maryland ore could not be treated efficiently by the flotation process.

Independently of the above experiments, samples of ore were submitted to other companies to determine if the Idaho Maryland ore would respond favorably to ball-mill crushing and flotation. A report was received from the Minerals Separation Co. stating that they had been successful in working out the problems with the ore treatment. They indicated that the tailing loss could be greatly reduced by ball-mill crushing, followed by flotation.

The footwall crosscut on the 1000-foot level was driven over 900 feet through altered diabase and gabbro before entering serpentine. The crosscut was stopped when no veins or fissures were discovered at that distance.

The east drift on the 1000-foot level was set up for electric trains and was prepared as a main haulage level for future ore production. This meant widening the drift in some places, laying 30-pound rail, and installing larger air and water lines. The drift extended 2,525 feet from the Idaho shaft and required retimbering part of that distance. This drift was readied for future exploration and development of the upward extension of the Dorsey and Dorsey footwall veins.

The drift west of the Idaho shaft on the 2000-foot level was extended 134 feet through an altered zone of slate with a heavy goudge and quartz stringers. Drifts were run both north and south along the contact zone. The north drift followed along a carbonaceous slate contact and quartz stringers for 60 feet. The south drift was driven 185 feet along the same altered-slate contact that had more quartz and better sulphides. These drifts were stopped when a 50 gpm water flow began.

The No. 1740 crosscut was driven 100 feet, and at a point 140 feet in the footwall of the No. 80 vein (or McGuire stope) a cross vein was encountered. A drift was driven 163 feet along this fracture and a strong fissure from six inches to two feet of heavy mineralized quartz appeared. The vein split beyond that point.

A raise was started near the intersection of the No. 1740 crosscut and the east-dipping No. 1740 vein. Ten feet up in the raise the vein was six feet between walls, with four feet of quartz showing gold, heavy sulphides and strong quartz stringers. It was necessary to stop this raise due to a more pressing need for miners in other headings.

The No. 89 west raise was started in the latter part of April, just west of the point where the 2000-foot level main footwall crosscut intersected the footwall veins. This raise was started to open up the complicated intersection of footwall veins developed on the 1600- and 1900-foot levels. This vein carried three feet of quartz, showing quartz stringers and considerable pyrite, with specks of gold and galena. The development ore was sent to the mill.

The No. 92 west drift was driven west from the No. 92 crosscut on the 2000-foot level. This drift was only advanced 50 feet before it was stopped. The vein showed 6 to 8 feet of formation, with coarse gold in quartz stringers. The drift was discontinued to use the miners in other faces.

After the company took over the Mike McGuire tribute, it was necessary to do development work in the stope for more efficient stoping. Intermediate drifts were run, and stoping carried on between drifts. The vein material in this stope was 6 to 8 feet in width, with 2 to 4 feet of quartz on the hanging wall, then a mass of quartz stringers in altered diabase, and more quartz on the serpentine footwall.

All of the material between walls was sent to the mill. Specimen gold was present periodically in the quartz. The ore averaged between $10 and $12 per ton, not counting the value of the sulphides.

The Metals Exploration Co. had started a raise from the 2000-foot level to connect with the bottom of the Dorsey winze. The first 50 feet was in good ore, but above this point serpentine was encountered and the vein pinched out along a steep wall. The raise was then abandoned.

The ore body in the footwall of this former raise, and the opening up of big veins in similar conditions on the Dorsey 1900-foot level, proved that the so-called pinching of veins and ore bodies against a steep wall on their upward course does not mean the termination of the ore body. In most cases the upward continuation of the vein and ore body could be found by crosscutting through the break at the H intersection.

The first work undertaken by the company after taking over the Yonker stope was to drive a crosscut into the footwall of the No. 1950 vein from the west end of the Yonker group's intermediate drift. This was located 60 feet vertically above the 1900-foot level.

At a point 50 feet in the footwall this new crosscut struck 12" of quartz, and six feet farther another 18" of quartz was cut. A sheltered mass of highly mineralized diabase, with numerous quartz stringers, galena and fine gold was found between these two veins. A drift was driven along the Dorsey footwall vein, with heavy gold, galena and chalcopyrite showing in every round.

Ore production decreased during the latter part of 1927, which was attributed to several factors. During this period it was necessary to place entire dependence on the No. 80 stope for company ore. A barren zone within the stope reduced production. During this same period the tributers found it necessary to do additional development work, which also affected production. The bullion sales for 1927 were valued at $91,328, and the concentrates for a total of $109,088.

Underground Operations—1928

In the latter part of 1927 Errol MacBoyle and Bert Crase examined an old crosscut driven from the footwall of the No. 1950 vein on the 1900-foot level of the Dorsey winze. A decision was made to extend the crosscut, and the first round exposed a well-defined quartz vein two feet in width, showing wire gold, galena and chalcopyrite. This vein was referred to as the No. 89 vein.

The crosscut was driven another ten feet and on January 1, 1928, a new vein showing from six to eight feet of ribbon quartz was cut. This new footwall vein showed galena, iron pyrite, and chalcopyrite with considerable fine gold. In the words of MacBoyle, "It is one of the strongest and best looking veins ever opened in the Idaho Maryland or other mines in the district."

The same crosscut had been driven several years earlier by the John A. Fulton regime and then stopped. It was extended in 1924 under M. A. Roche, but once again it was stopped—just one round short of exposing the footwall veins.

The new vein was designated as the No. 1900 vein. It was the same Dorsey footwall vein that had been stoped by the Fulton regime at a point 100 feet east of the present discovery. A raise had been put up 91 feet at that time, and the vein averaged five feet in width, with values from $12 to $13 per ton. This work ceased when the Dorsey winze area was abandoned in 1924. A raise was to be put up from the 2000-foot level to mine the vein, but this work did not happen.

All work on company account was stopped in other areas of the mine to concentrate on exploring the Dorsey footwall veins. Priority was given to developing the newly discovered No. 1900 vein. During 1928 it was developed for 700 feet on the 1900-foot level. Two raises were put up on the vein for a distance of 180 feet each. At a point 75 feet above the 1900-foot level an intermediate drift was driven 300 feet.

In both drifts and raises this vein ranged in width from 5 to 10 feet, with 3 to 8 feet of ribbon quartz containing iron pyrite, chalcopyrite, galena and visible gold nearly every round. Heavy specimen gold was found at frequent intervals in the course of the development work.

At a point 30 feet above the 1900-foot level in No. 1 raise, the No. 1900 vein was cut off by a large fault dipping 70° to 80° south. This was the same fault encountered in the Yonker group's intermediate level, above and 100 feet to the west. The raise was continued up through the fault, and at 60 feet above the level the quartz came back in. The vein widened with each subsequent round, showing heavier gold and galena.

Progress in opening the Dorsey footwall veins had been slower than anticipated, due to the faulted and broken zones resulting from crossings and intersections. Much of the work was done at a point where the No. 1950 (or hanging-wall) vein, the so-called No. 89 vein, and the fault and Dorsey footwall veins approached one another on their upward courses. It became necessary to heavily timber all the drifts and raises, and consequently progress had been slowed.

Tributers Operations—1928

During the year the number of tributers or leasers was reduced to twelve men. The work was confined to the development and stoping of the Morehouse vein.

The Stambaugh miners concentrated on stoping the Morehouse vein on the 2300-foot level in No. 45 winze. The stoping widths ranged between 6 and 10 feet. Heavy gold was found in the quartz stringers much of the time, and specimen gold occasionally. The mill runs showed a recovered value of $12 to $14 per ton, even with a large amount of waste rock included. The two stopes produced well and kept five stamps in continuous operation.

The William Harvey group continued stoping on the Morehouse vein on the 2350-foot level. The vein was well mineralized and ranged between 2 to 4 feet in width. The ore averaged between $8 and $10 per ton. Heavy specimen gold occurred frequently.

Milling Operations—1928

The ore milled during 1928 came mostly from development drifts and raises, and contained from 50 to 75 percent waste rock. Owing to underground operating conditions and the lack of a sorting plant, it was impossible to segregate even barren crosscut waste from going to the mill.

Management was anxious to increase production based upon the ore showing underground. This could not happen, due to the present mill capacity. The decision then had to be made whether to double the size of the stamp mill or to treat part of the ore in the New Brunswick mill. The intention was to gradually increase the treated ore to 10,000 tons per month.

Surface Operations—1928

In July the front legs of the headframe began to settle, and by mid-September it became imperative that repair work be done at once. The headframe had reached a point where it was eleven inches out of line, and the cable began to come off the sheave wheels. The legs were settling because the headframe had been placed on timbers set on fill ground instead of solid concrete foundations.

The mill and underground operations were suspended until this and other necessary repairs were completed. The entire crew of underground, surface and mill men was retained to expedite the work. The items completed were: repairs to the hoist, pouring concrete pillars, realignment of the headframe, placing concrete retaining walls around the shaft collar, replacing battery posts in the mill, and repairing shaft timbers down to the drain tunnel.

Production—1928

During 1928 there were 15,557 tons of ore milled, from which $87,059 was recovered by amalgamation and $20,759 by concentration. The total recovery was $107,810, or $6.93 per ton milled. It was estimated that over $2 per ton was lost in the tailings, a good part of which could have been recovered if the mill were adequately equipped with concentrators and had a cyanide plant. The gross value per ton milled was over $8.93 per ton. This was an excellent showing, considering the dilution.

CHAPTER 9: REORGANIZATION OF IDAHO MARYLAND CONSOLIDATED

In January 1929 new financing was sought to make possible further development of the rich No. 1900 vein, explore for other ore bodies, expand milling facilities and reopen the Brunswick mine. Errol MacBoyle spent a lot of time away from the mine on trips to New York and San Francisco to seek outside capital. Money was tight at the time, and he was unable to obtain new financing.

It then became necessary to sell stock in the corporation to raise the needed funding to carry out these objectives. The California Corporation Commission gave Idaho Maryland Consolidated Mines, Inc. permission to sell $500,000 worth of common stock. The shares became available to the public for the first time on April 16, 1929, when the stock was listed on the San Francisco Mining Exchange.

Idaho Maryland Consolidated Mines was capitalized for 2,500,000 shares of common stock, of which 1,541,932 had been issued, and 500,000 shares of preferred stock, of which 220,966 were outstanding. Only the common shares were listed on the exchange.

During 1929 a contract was entered into with Ackerman and Co., San Francisco brokers, whereby they were to purchase 500,000 shares of common stock for $500,000. Unfortunately, the stock market crashed in October, triggering the great depression of the 1930s. After $110,000 had been paid into the treasury under the agreement, the brokers were unable to carry out their contract. The result was that the plans for development of the properties had to be changed, and work was confined to the No. 1900 ore body.

The corporate directorate was reorganized to include three new members. The board was made up of Edwin L. Oliver, president; Frederick McNear, vice president; Errol MacBoyle, managing director; and Paul Claiborne, secretary and treasurer. The directors included L. H. Duchak, W. E. Casson, Joseph Smyth, William Muller and L. H. Peters.

Dual Fatalities—1929

On April 15, 1929, John F. Goreman was transferring ore along the overhead trestle from the headframe ore bin to the ore bin at the crusher building. The drive wheels on the trolley motor lost traction on the slight incline and the train stopped.

Raymond Harris, the crusher operator, offered to help Goreman clean the film of oil from the steel rails and the drive wheels. The train was backed up onto the seldom used east trestle which led from the headframe waste bin to the waste dump.

The two men began wiping the oil from the wheels when a section of the 30-foot-high trestle suddenly collapsed, hurling Goreman and Harris to their instant deaths amid a mass of ore, steel cars and broken timbers.

Hoistman Leroy Bond gave five blasts on the mine whistle, which brought many surface workers to the accident scene to extricate the bodies. Both men had expired before Dr. Carl P. Jones arrived at the mine.

Underground—1929

All of the tributes or leases were discontinued in February, with the exception of the Stambaugh lease on the 2300-foot level in the No. 45 winze, and the Harvey lease on the 2350-foot level in the No. 87 winze. Both leases were producing good ore, ranging from $10 to $15 per ton.

The company concentrated on developing the No. 1900 vein during 1929. It was explored for over 800 feet laterally on the 1900-foot level, and by three raises extending over 250 feet above that level. Drifts were run for over 400 feet on both the 1800- and 1700-foot levels. Near the end of the year the 1600-foot level was started with 48 feet of drifting. The development on the No. 1900 vein at this point disclosed an immense tonnage of ore that was virgin to the surface.

The quartz ranged from 4 to 8 feet on the 1900-foot level and increased in width on the levels above. The quartz was heavily ribboned throughout, with iron pyrites, galena and chalcopyrite showing fine gold nearly every round. Specimen gold was present periodically on all four levels and in the raises.

All of the development rock from the drifts and raises on the No. 1900 vein had to be sent to the mill as ore. There was no place underground to sort out the waste, and the rock was too good to send to the waste dump. There was at least 50 percent waste dilution. Two stopes were opened on the 1900-foot level to send better grade of ore to the mill.

Development on the 1500-foot level west of the main shaft discovered the Morehouse vein. This lined up well with the Morehouse workings on the 2000-foot level. The vein carried two feet of quartz and heavy specimen gold at the point of discovery. The Morehouse vein was believed to be one of the main fissures of the area.

It became necessary to rebuild the wooden headframe, due to its age and weakened condition. It had been used at the Union Hill mine, then dismantled in 1920 and moved to the Idaho shaft by the Metals Exploration Co. Many timbers showed damage from cracking and dry rot.

Underground mining was discontinued for nearly a month, beginning in the latter part of September, while the headframe repairs were underway. The underground pumps were operated daily during the shutdown to keep the lower levels free of water. The pumpmen and a small crew of timbermen had to use the shaft ladderway to reach the 1000-foot level. The men were then able to be lowered down the Canyon shaft to the 2000-foot level.

Milling Operations—1929

The stamp mill operated at capacity and could not handle any increased production from the ore already blocked out on the No. 1900 vein. A decision had to be made whether to build a larger facility by adding 20 more stamps, or to build a more costly mill by introducing a ball mill and flotation process.

Because of its complexity, several metallurgists were brought in to test the ore to determine the best milling process to follow. Exhaustive studies and tests were conducted. The results attained by these metallurgists were unsatisfactory. A decision was made to try classification and gravity concentration before deciding to install either flotation or cyanide treatment.

Several changes were made in the mill to increase the efficiency of the metallurgical treatment. A classifier and dewatering cones were added to the flow sheet. The sand product from the classifier was sent to the Deister Plate tables and the middlings from these tables were pumped to the Frue vanners. The slimes from the classifier were also pumped to the Frue vanners. It was hoped that the double treatment would lower the tailing loss and result in saving a larger percentage of sulphides, especially the very high-grade sulphide slimes.

The concentrates had an average gross value of $160.00 per ton, or $1.95 per ton milled, which were the highest grade concentrates produced in recent years. The concentrates showed a marked increase in value after the No. 1900 vein ore was milled.

There were 17,241 tons milled during 1929, with a recovery of $155,876 for an average of $9.04 per ton. The average tailing loss was $2 per ton, showing a gross value over $11 per ton milled. This was a remarkable showing, considering there was 227 lineal feet of crosscut waste included in the rock milled.

Worker's Compensation Insurance—1930

The California Inspection Rating Bureau increased the worker's compensation rate for the mining industry from $4.95 to $10.50 per $100.00 of payroll. This was a tremendous increase in overhead cost. In order to reduce this high cost, the corporate directors opted to self-insure for worker's compensation insurance.

The California Accident Commission granted the corporation permission to insure itself up to $5,000. A required $50,000 surety bond, to be in force for ten years beyond the life of the mine, was secured from Fidelity and Casualty Co. A reserve fund for paying workers' claims and expenses was established. A contract was signed with Dr. Carl P. Jones of Grass Valley for medical and hospital care for injured employees.

The corporation also obtained a policy with Lloyds of London, insuring the company against excess liability for worker's compensation caused by an accident or catastrophe in excess of $5,000 and up to $200,000.

During 1930 the company paid a $241,000 obligation bond owed to the Harry Paine Whitney estate. As a consequence of this move, all notes, bonds, etc. of the operating subsidiaries were now entirely owned by the holding corporation. This included 72 percent of the stock of the Idaho Maryland Mines Co., 99 percent of the Brunswick Consolidated Gold Mining Co. and 100 percent of the Morehouse mine.

Underground Development—1930

During 1929 a crosscut had been driven in the footwall of the east drift on the 1000-foot level to intersect the upper extension of the No. 1900 ore body and the Dorsey footwall veins. This crosscut had been stopped due to lack of development funds. The crosscut was started again in 1930, and by the end of the year it had reached 520 feet without cutting any veins.

Two raises were run up in line with each other—one from the 2000-foot level to the 1950-foot level, and the other from the 1950-foot level to the 1900-foot level. These were run in line with the existing No. 3 raise, which then provided a main working raise for the No. 1900 ore body from the 2000-foot level to the 1500-foot level. The No. 3 raise greatly facilitated the movement of men, ore and supplies, thus reducing the cost per ton of ore. The broken rock from the 1500-foot level and all levels below could now be dropped directly to the 2000-foot level for tramming to the main shaft by electric trolley trains.

A new 3-ton double-motored electric locomotive, made by mine mechanics at a cost of $500, was placed in operation on the 2000-foot level. This trolley locomotive pulled twenty loaded ton-cars, as compared with ten cars hauled by the smaller motor.

The No. 4 vein split off the No. 1900 vein into the footwall and was encountered on the 1900-, 1800-, 1600- and 1500-foot levels. Each split showed from three to four feet of ribboned quartz with galena, chalcopyrite and fine gold.

Results of the development work on the No. 1900 vein were highly satisfactory. This block of ore had now been proven for a length of 1,000 feet on the 1900-foot level. Intermediate drifts were driven on the 1950-, 1900-, 1800-, 1700-, 1600- and 1500-foot levels. Raises were run up at intervals to block out the ore body for future stoping.

The vein ranged in width from 3 to 15 feet between walls. The quartz was ribboned with galena, iron pyrites, chalcopyrite and free gold on each level. The average width of the vein was six feet, and the gold values increased in the upward phase of development. Pockets of heavy specimen gold were found periodically on each level. Considerable specimen gold was found on the 1600-foot level and in the No. 3 raise above the 1500-foot level.

Milling Operations—1930

Milling operations continued to be a problem. Experiments to re-

Mantruck for transporting men and supplies in the No. 3 raise.

New Brunswick headframe, dry building and shop buildings, 1930.

New Brunswick mine, showing mill, headframe, and truck haulage ramp at far right.

duce the tailing losses were unsuccessful, and production was limited by having only 20 stamps. Insufficient milling capacity began to create restraints on underground ore production, and this kept the overhead cost to approximately $10 per ton.

In October 1930 the company began modifying the Brunswick mill in order to double ore production. William H. Sheeler, mill expert from Silverton, British Columbia, was placed in charge of the new equipment and metallurgical work. He installed a flotation process (designed by himself) to be used in conjunction with the existing 20 stamps.

Use of the Brunswick mill meant hauling the rock between the two mines. First an ore bin had to be constructed at the Idaho Maryland for loading ore into trucks. A truck ramp and ore bins were also constructed at the Brunswick mill to receive the ore. The mine road had to be improved to accommodate truck travel. A contract was entered into with Luther Courtright to haul the ore to the Brunswick mill at a rate of 25 cents per ton.

During 1930 there were 19,451 tons treated in the Idaho Maryland stamp mill, with a gross recovery of $241,059, or $12.40 per ton. Owing to inadequate and out-of-date mill equipment, the gold loss in the tailings averaged $2.04 per ton, making a gross value of the ore for the year $14.45 per ton. The net production, after deducting bullion and smelter charges, was $234,752. The payroll increased during 1930 from 90 employees to 130.

Milling Operations—1931

The Brunswick 20-stamp mill began operation on January 2 with marginal success. Even after continued experimentation with the new flotation process, the tailing loss was too high. The results obtained by Mr. Sheeler were unsatisfactory, and it was decided to dispense with his services on May 25.

Pan-American Engineering Corp. had worked out a treatment of the Idaho Maryland ore which that showed in the laboratory a tailing loss of only 60 cents, as compared with a tailing loss by Sheeler of from $1.20 to $1.50 per ton. Pan-American guaranteed a loss of 40 cents if the ore was ground to 80 mesh. They also guaranteed a much higher grade of concentrate.

The Kraut flotation cells were perfected to handle this specific type of ore, and especially the talc in the ore.

During the latter part of 1931 concrete foundations were constructed in the Brunswick mill for the installation of a 7-foot by 5½-foot ball mill, Dorr classifier and the Kraut flotation cells. Several changes had to be made to accommodate the new equipment. These changes were made without affecting normal milling operations.

The Idaho Maryland mill was enlarged to house a change room, a room for retorting and melting gold, and a canvas plant to save values being lost in the tailings, and a bank of Kraut flotation cells.

Underground Operations—1931

The vein-numbering system was changed during 1931. The original Eureka-Idaho-Maryland vein was changed to the No. 1 vein, the Dorsey vein was changed to the No. 2 vein, the No. 1900 vein was changed to the No. 3 vein, and the new footwall branch from the No. 1900 vein became the No. 4 vein.

The footwall crosscut on the 1000-foot level was stopped at 945 feet from the east drift. It became evident that the crosscut had been driven too far without encountering the No. 3 ore body or the Dorsey footwall veins. The decision was then made to extend the east drift farther east and drive a shorter crosscut to intersect the veins.

Development operations were confined to the No. 3 ore shoot, because it gave a greater chance for immediate and continued profit than from any other part of the mine. The intermediate drifts from the No. 3 raise continued to be developed both east and west, and the No. 3 and No. 4 veins provided good mill rock. The No. 3 raise had reached a point 200 feet above the 1500-foot level, and the 1400-foot intermediate level was established at the 180-foot mark above the 1500-foot level. Several stopes were started to provide cleaner ore, to offset the inclusion of waste from the development headings.

A hoist was installed on the 1500-foot level for hoisting men and supplies in the No. 3 raise. This was very cost effective, as men arrived at their work places earlier, and timber handling was greatly reduced. All the drifts in this area were heavily timbered, to withstand the swelling ground associated with the serpentine rock. Timber crews were kept busy replacing broken timber as it occurred.

On the afternoon of December 1, 1931, Cecil Norris was transferring ore from one chute into another on the 1500-foot level. He was standing on a wooden platform when a large rock struck one of its supports, causing the platform to fall into the chute. Norris fell approximately 60 feet and was killed instantly. His partner, George Beck, was able to stop the flow of rock at once and then summon help. The body was brought to the surface at 2:00 p.m. and Dr. Carl Jones said Norris's skull had been fractured and he had suffered many bone fractures and body bruises.

As of December 31, 1931, there was no outstanding indebtedness. Payments of dividends on 220,996 preferred (8%) shares had been made. Provisions under which preferred stock was issued required its complete retirement before dividends could be paid on common stock. It appeared probable that current production would be ample to retire all preferred stock during 1932 and still have sufficient working capital.

During 1931 there were 54,375 tons of ore produced, 30,894 tons of which had been milled at the Brunswick plant and 23,481 tons at the Idaho Maryland plant. The gross gold and silver recovered from 54,375 tons milled was $752,657, or an average of $13.84 per ton milled. The average loss in tailings for the year at both plants was $2.13 per ton. Recent changes in metallurgical practice and new equipment at the Brunswick mill resulted in a tailing loss of only 55 cents.

The tailing loss, added to the gross recovery, gave a gross value of $15.97 per ton of ore milled for the year, notwithstanding the large percentage of development ore milled, which tended to dilute the higher grade of ore from the stopes. The net production, after deducting bullion and smelter charges, was $741,311.

Total expenditures were $401,083, which included development, overhead, worker's compensation insurance, taxes, metallurgical experiments, and the cost of operations. The total cost per ton milled was $7.38.

Surface Work—1932

The corporation made great strides in 1932 toward upgrading the surface works. A 30-foot-high semi-circular-shaped dam was constructed to form a large lake for impounding mill tailings. The dam was built with waste rock hauled over from the main waste dump. An electric locomotive and a train of ore cars was used to move the material. The dam was located in the western part of the property, between the office and the Morehouse mine.

A timber yard was established in the large flat area between the new dam and Wolf Creek. The Nevada County Narrow Gauge railroad ran a spur track into the timber storage yard, where timber received on flat cars was off-loaded. Two garages, each with twenty car spaces, were constructed to protect employee cars from the elements. These were shed-type garages with full enclosure except for the drive-in openings. A separate garage was built behind the office for the company trucks.

The Cahill Construction Co. from San Francisco was awarded a $6,700 contract to design and erect a new 105-foot-high steel headframe. First, company carpenters and labor crews were assigned to complete preparatory work, such as relocating shop buildings and constructing twelve large concrete footings to support the heavy headframe girders.

Some of the mine's past history came to light during the excavation phase for the pillars. A red brick foundation was uncovered, which was the remains of a kiln used to generate carbonic acid to suppress a mine fire on the 1000-foot level in 1889. Ashes and burned wood debris were found that dated back to 1893, when the headframe, hoisting works and many surface buildings were destroyed by fire. Blocks of Rocklin granite, used decades earlier for support of previous headframes, were also encountered. Pneumatic drills had to be used to cut through the granite.

The Cahill crews began erecting the new headframe during the first week of October, while dismantling the old wooden headframe at the same time.

A 200-ton-capacity ore and waste bin was constructed on the east side of the headframe. This structure was built with a ten-foot-high drive-through section beneath the bins, where company trucks could be loaded with either ore or waste rock from overhead chutes.

A 45,000-pound jaw crusher, along with an enclosed structure, was placed above the ore bin. Another ore and waste bin was built within the confines of the upper section of the headframe, just below the self-dumping skip rails. The two large sheave wheels were mounted on top of the headframe, in line with each hoisting compartment.

The ore, when hoisted, dumped automatically into the elevated ore bin, then fed through the jaw crusher into the lower ore bin, to be trucked to the mill. The waste rock, when hoisted, was dumped into the upper waste bin, then fed down into the lower waste bin and trucked to the waste dump.

When work was progressing on the headframe, a new hoisting works was also underway. Concrete foundations were prepared on ground hollowed out of the waste dump overlooking Wolf Creek. The 300 hp Weaman-Seaver-Morgan double-drum hoist was moved from the Brunswick mine and set in place.

Steel headframe being constructed over existing wooden headframe.

Bird's-eye view of the office, sand dam, timber storage, and employee garages next to Wolf Creek. Idaho Maryland Road is at far right.

Underground mining and milling operations were shut down for three weeks while the surface construction took place. The company resumed normal work on November 1, 1932, when 200 men were called back to work. Within two days the surface ore bins were full and both mills in operation.

The Nevada County Narrow Gauge Railroad built a spur track along the east side of the headframe. Supplies then could be off-loaded directly to the shaft collar by the derrick crane.

Underground—1932

Many additions and improvements were carried out underground during the year. The 1000- and 2000-foot main haulage levels were greatly improved for more rapid transportation. The main shaft was deepened to make room for four-ton skips, which replaced the former two-ton skips. The ore pocket on the 1000-foot level was enlarged to accommodate the anticipated increase of ore from that level.

The two east compartments of the main shaft were retimbered from the surface to the 300-foot level, and the west compartment was retimbered to the 150-foot level. Port Orford cedar was used in this operation because of its longer life in a dry atmosphere.

A duplicate emergency pumping system was installed with centrifugal pumps at the 700- and 1000-foot levels. The 700-foot level was reclaimed and repaired so that the large volume of water coming from the Eureka workings could be pumped to the surface from this level, instead of allowing the water to drop to the 1000-foot level pumps.

The development work on the 1000-foot level encountered the upper extension of the No. 3 ore body in May. The vein measured 3 to 10 feet in width, with good mineralization and free gold showing at frequent intervals. The discovery of good ore at this elevation strongly indicated that the ore body would continue upward, close to the surface. Increased production during the last half of the year was due in part to the ore produced from the drift and raises on the 1000-foot level.

The main No. 3 raise driven in the No. 3 ore shoot reached a point 30 feet above the 1300-foot level, and showed 15 feet of good ore. The No. 2 raise, 400 feet west of the No. 3 raise, was also in good ore. The only disappointing feature of the year had been the failure to develop the upward continuation of the No. 4 and No. 5 ore bodies. Further exploratory work on these two veins had to be postponed until operating conditions would permit economic handling of waste rock.

During 1932 approximately 5,000 feet of development work was done in the form of drifts, crosscuts and raises. Approximately one-third of the tonnage milled was produced from development work. The balance was mined from stopes.

Milling Operations—1932

At the Idaho Maryland mill a new metallurgical laboratory, with an electric furnace and complete experimental equipment, was installed. New ore bins and a ramp were constructed for receiving ore by truck haulage. The building was enlarged to include a flotation and concentration section. The main mill building was extended to give additional room for drying and storage of concentrates.

The ball mill and classifier were placed in operation on February 1 at the Brunswick mill. This gave much finer grinding and allowed the oversize sands to be pumped back to the ball mill for regrinding. A thickener for flotation concentrates was installed to work in conjunction with an Oliver filter. The long amalgamating plates were removed and four Deister amalgamation tables were substituted. These machines concentrated out the gold and rich telluride and galena concentrates, replacing amalgamation in the stamp batteries and on the plates. The 40-mesh battery screens used for amalgamation were replaced by 8-mesh screens, which stepped up tonnage from 80 tons to 120 tons per day.

Fatal Injury—1932

At approximately 2:45 p.m. on December 8, 1932, Frank Bennett was standing on a ladder, and he had just handed a piece of lumber to a fellow carpenter on the roof of the hoist building. Somehow he came in contact with a nearby 4,000-volt electric wire and fell about 10 feet to the ground. He was found several minutes later, lying on the ground in front of the ladder.

He had received severe electrical burns on his left arm and hand, and the tips of three fingers on his left hand, as well as serious head injuries from the fall. Bennett was treated at the scene by Dr. Carl P. Jones and transported to Jones Memorial Hospital by the Jefford Ambulance Service. He died that evening in the hospital without regaining consciousness.

On May 27, 1932, a fire of undetermined origin destroyed the surface plant at the Old Brunswick mine. This involved the hoisting works, compressor room, headframe and office building. This portion of the mine was not in use, and did not affect current operations.

The retirement of all outstanding preferred stock was completed on December 23, 1932, and all dividends were paid. The company had no other capital liability other than the 1,701,932 shares of common stock outstanding.

Production—1932

During 1932 there were 58,245 tons of ore produced, 36,805 of which were milled at the New Brunswick plant, and 21,440 tons at the Idaho Maryland mill. The gross gold and silver recovered from both mills was $449,470, for an average of $16.82 per ton milled. The average loss in the tailings for the year at both plants was $1.14 per ton. The tailing loss added to the gross recovery gave a gross average value of $17.96 per ton. The net production, after deducting bullion and smelter charges, was $954,712. The large percentage of development ore that was milled tended to dilute the higher grade ore from stopes, otherwise the average grade would have been higher.

Total expenditures during 1932, excluding interest paid and smelter charges, amounted to $485,937, including compensation insurance, local and state taxes, mine development, metallurgical experiments, and costs of alterations at both mills. The total cost per ton milled was $8.34. The operating profit was $470,172 before federal taxes, depletion, depreciation or interest on debt owed to the Idaho Maryland Consolidated Mines, Inc.

Winter Storms Affect Operations—1933

The board of directors granted Errol MacBoyle a six-month leave of absence with full pay, beginning January 10, 1933. This was for his conscientious devotion to the interest of the corporation for six years without taking any vacation.

The mine was forced to cease operations for over a week, during the latter part of January, due to the lack of water to operated the mills and air compressors. The Cascade ditch had become clogged with ice and snow, cutting off water flow to the mine reservoir. At the same time, 4-foot to 6-foot snow drifts closed the road between the Idaho Maryland and Brunswick mill. Mine employees used a Caterpillar tractor to clear the road so truck haulage could resume.

Stanford University Geology Professor Clifford F. Tolman was elected to the corporation board of directors, along with Stuart S. Hawley. Professor Tolman was hired as a consulting geologist to plan future development work.

Double-drum hoist set in place.

Hoist building being constructed over hoisting works.

(Left)
Headframe complex under construction.

(Below)
Headframe, rail system, crusher building and ore bins completed.

A new 2800-cubic-foot Ingersoll Rand air compressor was purchased to meet the needs of increased production. The hoist room building was extended to house the new compressor, and a relocated compressor of the same size.

Development Work—1933

Consulting geologist Clifford Tolman's theory regarding development work was to drive three feet of development in waste for every one foot of development in ore. Therefore, he felt the company had a long way to go in catching up on the necessary exploratory work in waste, based upon several years of developing the No. 3 ore body. He outlined five proposed development projects that should begin in 1933:

> **A.** Crosscut on the 1500-foot level from the Idaho shaft to reach the continuation of the Idaho Maryland vein. If encountered, then drift on the structure to the Morehouse footwall vein to determine if the Idaho Maryland vein bends into the Morehouse footwall vein.
>
> **B.** Develop the Morehouse footwall vein on the 2000-foot level.
>
> **C.** Crosscut under the McCarty ranch property from the 1000-foot level, to explore the five vein-systems of the Pioneer mine.
>
> **D.** Sink the New Brunswick shaft to a depth of 5,000 feet.
>
> **E.** Connect the Idaho Maryland 2000-foot level with the New Brunswick shaft.

An ambitious development program began during 1933 following Tolman's recommendations. The No. 712 crosscut on the 1000-foot level, more commonly referred to as the Mitchell crosscut, was driven 1,459 feet under the McCarty ranch during 1933. The crosscut was driven through serpentine, with only two narrow pre-mineral structures encountered during 1933. The "L" fault was exposed soon after the crosscut was started.

The 1000-foot level workings had reached a distance of over 8,000 feet from the main shaft, and better air circulation became necessary. The area between the 1500-foot level of the No. 3 ore body and the 1000-foot level was also in dire need of better air circulation.

To solve this problem, a decision was made to sink a winze on the the dip of the No. 3 vein from the 1000-foot level down to the 1400-foot level. At that point the winze would connect with a raise previously run from the 1500-foot level. The No. 201 winze was started in June, and had advanced 393 feet by the end of the year.

Another winze was sunk a distance of 246 feet from the 1900-foot level down to the 2000-foot level. This winze was to explore the No. 5 vein. The winze was not extended beyond that point.

A long crosscut was run from the Idaho shaft station on the 1500-foot level to intersect the No. 1 vein. This vein was advanced over 1,000 feet, extending beyond the bottom of the Eureka shaft. No ore was found in that area of the mine. The No. 1 vein structure was quite narrow, with very little quartz present.

A drift 514 feet long was driven on the Morehouse footwall vein on the 2000-foot level. For a short distance two feet of quartz was showing, but it assayed only $2.48 per ton. The drift followed a goudge or seam in unmineralized serpentine for most of the distance.

Drifts were run both east and west from the main No. 3 raise on the 1300-, 1200- and 1000-foot levels within the No. 3 ore body. Drifting also took place on the No. 5, No. 6 and No. 7 veins east of the No. 3 raise. Drifts were driven on the No. 3 and No. 13 veins on the 900-, 850- and 800-foot levels, from raises run up from the 1000-foot level.

Reopening the Brunswick Mine—1933

On July 1, 1933, the Idaho Maryland Mines Co. leased the Brunswick mine from the Idaho Maryland Consolidated Mines, Inc. The lease was for five years, with a royalty of ten percent of the gross recoveries of gold and other precious metals, with a minimum royalty of $500 per month.

The long-awaited task of reopening the Brunswick mine started on August 15. The first work began at the Old Brunswick mine site, where it was necessary to clear away debris left from a fire that destroyed the hoisting works two years earlier.

Temporary hoisting equipment was set up to recapture the caved incline shaft. Crews of miners worked three shifts daily, six days a week, to remove the caved ground and old timber while retimbering the three shaft compartments. It took eight weeks to reopen the shaft, which was completely caved for 190 feet.

Millright Robert Hathaway, with a crew of carpenters and laborers, rebuilt the wooden headframe, ore bins, hoist room and compressor house, while the shaft work was underway. The double-drum hoist was brought over from the Idaho Maryland for use at this shaft.

Work was also underway at the New Brunswick shaft in preparation for removing water from the mine. A 600 hp double-drum hoist was purchased from the Allison Ranch mine and installed in the Brunswick hoist room. Large bailing skips were used for bailing water from the shaft. An air-lift system was also installed. A large wooden trough was installed above the shaft collar to divert the water to Little Wolf Creek.

Bailing water started on September 26, and the air lift began the following day. Water was removed at a rate of between 720 and 800 gpm by the two methods together. Both shafts are connected underground by a drift on the 900-foot level, therefore the water receded in both shafts simultaneously. These two methods of dewatering the mine proved satisfactory, and by the end of the year the water level was below 950 feet.

New Mill Construction—1933

Fifty-five tons of concentrates were shipped to the Amador Metal Reduction Co. to be treated in their cyanide plant. This was to obtain a test run on a substantial quantity of concentrates, in order to assist management in its determination of the best method of constructing a new cyanide plant.

The final step in modernizing the Idaho Maryland surface plant began in mid-November with a crew of 25 carpenters under the guidance of Bob Hathaway. The new mill building construction began on the sloping hillside behind the stamp mill. The project also included a cyanide plant adjacent to the new mill building.

The milling process was to include a ball mill, classifier, and flotation cells, with ample room for additional equipment. Plans were to return several years' accumulation of mill tailings through the mill. The tailings were believed to average around $3 per ton, and were regarded as a potential source of income.

The cyanide process was designed to recover the gold values from the concentrates generated at the Brunswick and Idaho Maryland mills. This would add an extra step in the gold recovery, but would eliminate the cost of shipping and smelting the concentrates at the Selby smelter.

Production—1933

Errol MacBoyle reported a satisfactory year from the standpoint of underground development, ore reserves, production and profits. The outlook for the coming year was very favorable.

The consolidated net income for the Idaho Maryland Consolidated

Old Brunswick mine headframe, hoist room and compressor room.

Hoist room, headframe and trestle of the Bullion mine south of Grass Valley on old Highway 49 (now La Barr Meadows Road).

Mines, Inc., and its subsidiary, Idaho Maryland Mines Co., was $359,072 for the year 1933. This was after reserve for taxes, but before provisions for depreciation, depletion and amortization charges. Dividends were 21 cents a share on 1,761,932 shares outstanding.

The Idaho Maryland Mines Co. produced 68,233 tons of ore during 1933. This had a gross value of $913,226, or an average of $13.38 a ton. Expenses amounted to $586,396. Net income of the operating company, prior to interest charges, totaled $505,108. Interest charges on the old debt were $18 31,938. The net profit after interest amounted to $296,083. However, the operating company had debts in excess of $3,500,000 which it was paying off.

Increased production and a stepped-up development program required more miners to be hired. The payroll rose to 300 employees by the end of 1933.

Bullion Mine

In 1933 the Bullion mine was reopened, after being closed for 26 years. The Bullion Gold Mines, Inc. was headed by Frederick McNear, president; Charles Brockington, manager; and Errol MacBoyle, consulting engineer. The mine was located on the east side of the Auburn highway, two miles south of Grass Valley.

Dewatering the shaft began on May 4, 1933, while at the same time buildings were being constructed on the surface. A combination machine and blacksmith shop was constructed on the east side of the highway, near the shaft collar. The headframe and ore bins were positioned close to the highway right-of-way on the west side of the highway. The hoist room, compressor room, and dry building were also constructed on the west side of the highway.

California Division of Highways engineer Charles Whitmore agreed to shift the highway right-of-way 12 feet to the west to facilitate the mine operation. Whitmore also gave permission for a trestle to be constructed over the highway, from the shaft collar to the headframe.

The mine was dewatered down to the 1500-foot level, which was the lowest level in the mine. The shaft was rehabilitated where necessary, and the existing levels inspected. Arrangements were made with the Idaho Maryland Consolidated Mines Corporation for the ore to be trucked to the Idaho Maryland mine and be milled in the old stamp mill. During the year twenty men worked underground and another twenty worked on the surface.

CHAPTER 10: GOLD PRICES RISE

On January 30, 1934, President Franklin Delano Roosevelt devaluated the dollar and gave the world a standing offer to buy gold according to the needs of international trade. He fixed the value of the dollar at 59.06 percent of its former equivalent, and on this basis a price of $35 an ounce was established for international dealings in gold.

Errol MacBoyle had every reason to be very pleased with the $14.33 increase in the price of gold—from $20.67 to $35.00 per ounce. This meant that several of the Idaho Maryland veins, previously deemed marginal, could now be mined at a profit. The longevity of the mines were insured because of the significant potential increase in ore reserves.

Pacific Gas and Electric Co. had been penalizing the mines whenever a sudden peak in power usage occurred. The peak rate was then charged for the next three months. A review of why these sudden peaks happened showed that the pumping hours were at fault. In order to eliminate these occurrences it was necessary to enlarge several underground sumps to allow for a minimum of 17 hours of water storage in each sump.

The power consumption was balanced out by hoisting rock on day and night shifts, operating the large pumps only on the graveyard shift, and running the mills on all three shifts. The power consumption of small float-controlled pumps did not affect the peak power problem.

Governor Frank B. Merriam revived the State Mining Board by

appointing five new members. The board was organized on August 16, 1934, while meeting with the governor. Errol MacBoyle was elected chairman of the new board.

New Mill and Cyanide Plant—1934

Work on the new mill and cyanide plant designed by metallurgist Roy Marcellus progressed satisfactorily during the winter months. The flotation portion became operational in May 1934. This section consisted of a 7-foot by 5-foot ball mill, a 55-foot-diameter thickener tank, an 8-foot by 20-foot Dorr bowl classifier, a bank of eight 36-inch Fagergren flotation cells, a bank of six Kraut flotation cells, an Oliver filter, and numerous Wilfly pumps.

Following completion of the flotation plant, mill tailings were pumped to the classifier from the tailings pond. The fines from the classifier were then pumped to the large outside thickener tank. The coarse material from the classifier was pumped to the ball mill for regrinding. The fines were fed from the thickener tank to the flotation cells for separation of concentrates. The concentrates were dewatered in the Oliver filter before being forwarded to the cyanide plant.

The cyanide plant became operational on March 22, and the concentrates from both mines were processed right away. The concentrates were first reground from 35 mesh to 100 mesh in a tube mill, using Danish pebbles as the grinding agent. The fine concentrates were then pumped to three large cyanide treating vats containing dilute solution of sodium cyanide. The cyanide dissolved the fine gold, allowing the other minerals to fall out as waste material. The gold precipitated out of solution by adding zinc dust. The zinc combined with the cyanide, displacing the gold, which settled out as finely divided particles.

The precipitate solutions were pumped to the refinery section at the conclusion of the cyanide process. Final treatment occurred in the refinery for separation, reduction and retorting, to emerge at last as the finished product—the gold brick.

Power for the milling operations and cyanide plant was supplied by three 400 hp transformers mounted on a concrete base adjacent to the building. Fifty-three electric motors were used to operate the equipment in this phase of operation.

The final phase of the mill construction program started in early August 1934. The mill building was extended to include a 750-ton-capacity ore bin, an 8-foot by 6-foot Marcy ball mill, a large Dorr

Inside the Idaho Maryland refinery.

Two retort furnaces for melting gold to make gold bricks.

Rows of 16-foot timbers for underground use at the Idaho Maryland mine. The mill and cyanide plant can be seen in the background.

classifier, riffles, concentrating tables, an amalgamating barrel, ore feeders, conveyor belts and pumps.

The brow of the hill above the mill was leveled and a large ramp constructed in conjunction with the ore bin partitions. The ore was trucked from the shaft ore bin to the mill, a distance of 0.9 miles.

The new ball-mill section tied in with the regrind ball mill and flotation plant to provide a very effective modern mill. The new 400-ton-per-day-capacity mill began operation on November 23, when Idaho Maryland ore was delivered to the new plant. Mill superintendent Hamilton Lewers was in charge of equipment installation and operations.

Round Hole Shaft—1934

By 1934 the underground workings had extended eastward over 8,000 feet from the main shaft. Therefore, a decision was made to sink another shaft closer to where the underground operations were in progress. A new shaft was necessary to provide better ventilation, a second exit, and to reduce the time in getting men to their respective work areas.

The location chosen for the new shaft was near the intersection of Brunswick and Idaho Maryland roads. This location presented problems from a geological standpoint inasmuch as the rock formations consisted of serpentine, goudge, ankerite and gabbro. A conventional shaft, consisting of drilling and blasting, in this type of ground would have required timber and a considerable cost to maintain.

Because this shaft would not be used to hoist ore or waste rock, the decision was made to sink a single compartment, using a 60-inch-diameter experimental core drill designed by J. Branner Newsome. This experimental drill was fabricated in the Idaho Maryland shop.

The drill consisted of two parts: the driving mechanism and a core barrel. These were assembled as a single unit that was lowered into the hole. The drive mechanism was assembled in a circular cab, which contained room for the drill-runner. The cab was kept from rotating by screw jacks, which were set against the rock wall.

The drill unit was driven by a 40 hp motor, with reduction gears to give a core-barrel speed of 60 rpm. The core barrel was 15 feet long, with a cutting shoe 1½ inches thick by 12 inches high attached to the lower end. The outside diameter of the shoe was 59 9⁄16 inches. The actual cutting was done by chilled shot. Several feet of water were maintained in the hole for cooling the cutter shoe.

In addition to the drilling units, there was a core-puller, consisting of

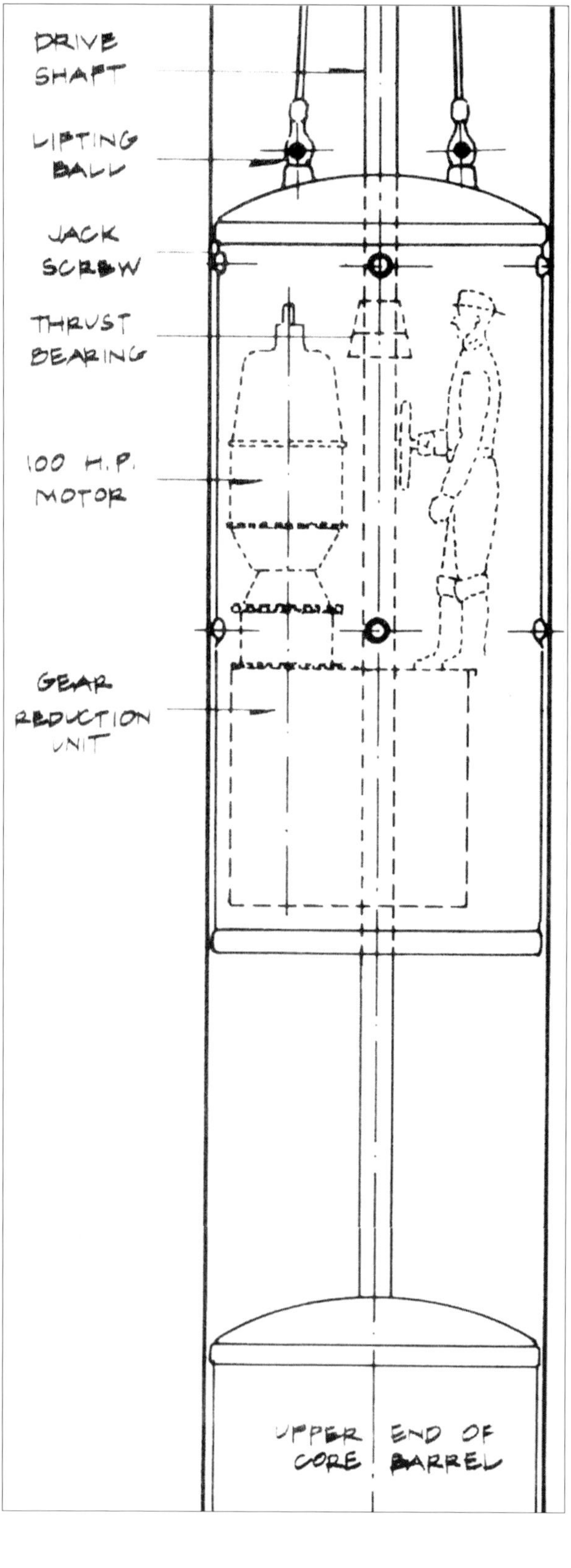

Engineer's drawing of J. Branner Newsome's revolutionary 60" diameter core drill. The drive mechanism was assembled in a circular cab, in which the operator stood. Chilled shot did the actual cutting.

a tapered ring and a number of fluted dogs. This was designed in such a manner that it could be lowered around a core, and when hoisting pressure was put on the puller the dogs would contract and grip the rock. After cores were removed there was always a certain amount of loose rock in the bottom of the hole. This was removed by shoveling it into an ordinary half-ton mucking bucket fitted with a safety hood.

Water was handled by an ordinary clack-valve bailing bucket holding about 150 gallons of water. When the drill was in the hole, men were raised or lowered in a specially designed container as a safety measure. Fresh air was furnished by running compressed air through a half-inch rubber hose. An ordinary bell system was used, with signals to fit the job.

The surface equipment included an 83-foot-high wooden headframe, consisting of 12-inch-square timbers. The hoist was a double-drum system with a 150 hp motor and a rope speed of 600 feet-per-minute. A small 20 hp hoist with built-up flanges to carry 1,150 feet of rubber-covered 3-conductor cable was used to raise and lower the power cable for the core drill.

The cores were hoisted above the shaft collar and a drawbridge placed over the shaft. The core was lowered onto a platform car and transferred to the waste dump. Some of the cores measured nine feet in length.

The No. 2 shaft, more commonly referred to as the "Round Hole" shaft, advanced 173 feet during 1934. The overall cost of sinking by this device had been approximately $15 per foot. The advantage of not blasting was apparent in that the shaft did not have to be timbered. The serpentine wall rock remained standing without any support.

Underground Accidents—1934

John Slavens and John Millikin were injured just before midnight on July 18, 1934. They were working in the east drift on the 1000-foot level when rock from the hanging wall suddenly fell, striking both men. At 1:30 a.m. the injured miners were brought to the surface, where Dr. Jones and the Jefford Ambulance Service awaited.

John Slavens received major head injuries, a compound fracture of his right arm and several contusions. He died in the Jones Memorial Hospital as Dr. Jones and Dr. John Loutzenheiser of San Francisco were trying to save his life in a series of emergency surgeries. John Millikin was treated at the hospital and allowed to go home to recover from lacerations and numerous contusions and abrasions.

On August 29 shift boss George Beck and mine superintendent Bert Crase were climbing up a raise above the Idaho Maryland 1000-foot level, when a falling rock swept them down the raise. Both men were brought to the surface in Stokes stretchers and taken to Jones Hospital by ambulance. They were treated by Dr. Carl Jones.

George Beck suffered a broken shoulder and contusions. Bert Crase sustained a broken right shoulder, four fractured ribs and a bruised leg, and he remained in the hospital for several weeks.

Shortly after the start of work on September 15, 1934, Everett C. Aslin was struck by falling rock in a stope on the 1300-foot level of the Brunswick mine. He suffered severe injuries, which included a fracture and dislocation of the right shoulder, dislocated right leg, fracture of the left leg, and internal injuries.

Aslin was treated by Dr. Jones at Jones Hospital. Because of the severity of his injuries, Dr. Thomas R. Hague, an orthopedic specialist, was flown to Grass Valley from Sacramento by MacBoyle's pilot, Latham A. Perrett. The two physicians worked together for several hours in a futile attempt to save the injured man. Everett Aslin died on the following morning at 9:30 a.m.

Development Work—1934

The development during the past year totaled 12,488 feet, from which 32,380 tons of ore were produced. Of this total 6,714 feet were driven at the Idaho Maryland mine, consisting of 2,672 feet of drifts, 2,976 feet of crosscuts, 769 feet of raises, and 285 feet of winzes.

The total amount expended in development at the Idaho Maryland mine was $243,147, representing a cost of $3.82 per ton milled. The average cost per foot of development amounted to $36.21 per foot or $4.99 per ton of rock broken in development. This work consisted principally in developing the block of No. 3 vein between the 1000- and 1400-foot levels. The rock was hoisted to the 1000-foot level through the newly completed No. 201 winze.

A raise was driven from the 850-foot level No. 13 drift, proving the upward continuation of the No. 3 vein along the No. 13 vein. Drifts were driven east and west from the 800-foot level station of the raise. The downward continuation of the No. 3 ore body below the 1900-foot level was proven by drifting on the No. 5 vein from a raise at the 1980-foot level elevation. The 712 (or Mitchell) crosscut had reached its objective and was stopped.

A total of 5,601 feet of development was driven in the Brunswick mine during 1934, from which 13,435 tons of ore was extracted. The total cost, including indirect charges, amounted to $115,788, or $7 per ton of ore milled. The total charges per foot were $20.67, including indirect charges. The total cost per ton of rock broken was $3.64.

The development consisted of 4,862 feet of drifting, 250 feet of crosscutting, and 489 feet of raising. The principal development consisted of outlining the ore shoots on the No. 1 vein. The 900-, 1100- and 1300-foot levels were driven on this vein. A crosscut was driven in the hanging wall to the No. 5 vein, and drifts were turned east and west on this vein on both the 1300- and 1100-foot levels.

The most encouraging development was the No. 2 ore body or "vertical" on the 1000-foot level. A drift was driven 600 feet on the flat No. 4 vein. The results of the Brunswick development program thus far had provided 300,000 tons of ore reserves.

Mill Operations—1934

The Brunswick stamp mill was operated solely on Idaho Maryland ore until July, when the Brunswick mine entered production. The mill continued to treat ore from both mines until the latter part of November, when the new Idaho Maryland mill began operations.

Due to the necessity to increase tonnage, the stamp battery screens were increased from 10 mesh to ⅜-inch and ½-inch openings; riffles were installed between the stamps and classifier, and a pump installed to pump the ball-mill discharge to the concentrating tables before returning to the classifier. The maximum tonnage treated with this installation was approximately 180 tons per 24 hours.

During 1934 a total of 80,237 tons of ore were milled from the two mines, with a gross recovered value of $1,472,902, or an average of $18.35 per ton. The total tailing losses were $1.25 per ton, making the average gross value $19.60 per ton. Idaho Maryland Consolidated Mines, Inc. made a net profit of $550,611 during 1934. This was before depreciation and depletion, and was equivalent to 31 cents on 1,726,932 shares of stock outstanding.

The dividends paid during 1934 amounted to 15 cents a share. A substantial amount was spent by the corporation for acquisition of additional Idaho Maryland Mines Co. stock. At the end of the year, all but one-half of one percent of the subsidiary had been acquired.

Bullion Mine—1934

Considerable work was carried out on the surface. Several men were hired to find the apex of the Bullion vein. A series of prospect shafts and trench excavations were dug over an area of considerable acreage. At least three shafts were equipped with windlass arrangements for working to a depth of 40 to 50 feet. Two other shafts were sent down for depths of 100 and 150 feet. These were equipped with compressed-air tuggers for hoisting. The payroll was increased temporarily to 104 employees.

Underground production during 1934 came from the 300-foot level, 800-foot level, and the north drift on the 1500-foot level. A ramp was constructed at the Idaho Maryland mine for trucks hauling Bullion ore to dump into the crusher building ore bins.

Management Simplified—1935

Great strides were made during 1935 in several areas of operation. Production was increased in both mines, milling facilities upgraded, and corporate management simplified.

Errol MacBoyle was appointed to the U.S. Bureau of Mines Advisory Board by Secretary of Interior Harold Ickes. Mr. and Mrs. MacBoyle flew to Washington D.C. in his low-wing Vulte monoplane to attend the meeting. The plane was piloted by Latham A. "Slim" Perrett.

On July 16, 1935, the Idaho Maryland Consolidated Mines, Inc. merged with its operating company, Idaho Maryland Mines Co. This move simplified the corporate organization under which properties were owned an operated, resulting in a more economical and efficient organization.

The name of the corporation was changed to Idaho Maryland Mines Corporation, and it was incorporated under the laws of the state of Nevada. The outstanding 1,739,432 shares continued to represent the assets of the corporation. The remaining shares of the Idaho Maryland Mines Co. stock were converted into shares of the consolidated corporation on the basis of one-fifth share of new stock for one share of the old.

In compliance with the California Mine Safety orders, mine employees were given first-aid training at the Veterans Memorial building in Grass Valley. Over 480 attended the classes during a three-week period. The training was conducted by Emory Smith from the U.S. Bureau of Mines office in Berkeley.

Ramp to crusher building to receive rock from Bullion mine.

Surface Improvements—1935

Surface improvements of the Idaho Maryland plant continued during the year. The headframe and several of the corrugated covered buildings were painted with an aluminum-colored paint to prevent rusting.

A 20-foot by 66-foot engineering office was built adjoining the main office building. The area beneath the engineering office was designed to be open on one side to accommodate eight car stalls for office personnel. A 10-foot-square concrete vault for protection of mine maps was constructed on the south side of the building, with a connection to the engineering room. The main office was renovated at the same time, to provide greater space for the executive offices and bookkeeping department.

The company contracted with the Fire Protection Engineering Co. to install sprinkler systems. This included the office buildings, the Idaho Maryland mill and the Brunswick mill. These were installed in conformance with specifications of the Board of Fire Underwriters. The main consideration to install these systems was for fire protection, and to guard against shut down and subsequent loss of production. The reduction in fire insurance premiums was also an incentive.

A new 45-foot by 50-foot automotive service and repair garage was built, complete with work benches and floor pits to service the modern fleet of pickups, trucks and construction equipment.

A power shovel was purchased for construction of a new proposed reservoir and other construction jobs. The shop mechanics constructed a boom to be used in place of the shovel bucket for unloading and stacking lumber.

Two garages for employees' cars were constructed on the flat area adjacent to Wolf Creek. The additional 35 spaces were badly needed for the men on the afternoon shift.

Underground Development—1935

Development in the Idaho Maryland during 1935 consisted mainly of extending the limits of the No. 3 and No. 13 ore bodies above the 1000-foot level. Blocks of ground on the No. 3 vein were prepared for future stoping. The limits of the No. 3 ore body had not as yet been defined, and exploration work continued with satisfactory results.

The Branner Newsome experimental boring drill holed through on the 600-foot level in July. Increased air circulation occurred immediately in all workings above the 1000-foot level following the connection. The

Drill core sections of serpentine taken during the sinking of the Idaho No. 2 "Round Hole" shaft at the Idaho Maryland mine.

Diamond driller Carl Hawke (left) and his helper, Albert Krasevag.

shaft had reached an elevation 960 feet below the surface by the end of the year.

The boring machine performed smoothly, except that in broken or soft ground it tended to get out of plumb. The harder and more uniform the ground, the better the drill performed. Cores up to nine feet in length were hoisted to the surface much of the time. Several mining men from the U.S. and foreign countries visited the site and showed interest in the boring machine's performance.

The diamond-drilling operations in the Brunswick mine were successful in discovering several new veins. Three new core drills were purchased to expand the drilling operations. The core-drilling program was placed under the supervision of the engineering department.

Sinking operations in the New Brunswick shaft began on March 1, 1935. Preliminary work involved sinking a single-compartment winze from the shaft bottom to the 1400-foot level. The winze was then enlarged to the shaft dimensions by raising up from the 1400-foot level. A large ore and waste pocket was run up to the 1300-foot level and grizzlies were installed over the pockets.

The 1400-foot level station was enlarged and a sinking hoist and a large-capacity waste bin provided for further shaft sinking. The shaft-sinking work progressed well, but was slowed by the preparation of the 1450-foot level ore and waste pockets.

Gust Seppala was seriously injured in a 50-foot fall in the Brunswick mine. The accident occurred on May 10, 1935, during the night shift. Seppala was brought to the surface in a stokes stretcher and taken by ambulance to the Jones Hospital. He had sustained a compound fracture of his left leg, pelvic and internal injuries. Gust Seppala died four days later from kidney paralysis caused by his injuries.

Development in the Brunswick consisted of drifting on the Nos. 1, 2 and 4 veins. Raises were put up on all three veins to block out ground for stoping and to assure adequate air circulation. Crosscuts were run on the 900- and 1300-foot levels to explore the Union Hill veins.

Investigation was carried out in regard to the most economical method to mine the Brunswick veins. Both shrinkage and open-stope methods were used first, but had to be abandoned due to the dilution of waste in the ore extracted. A combination of cut-and-fill and backfill stoping methods was devised, which materially reduced the dilution and increased the grade of ore. Costs had to be reduced to a point where the lowest grade of vein material could be mined at a profit, in order to insure minimum acceptable returns. There were 22,591 linear feet of development within the two mines during 1935, and 8,164 feet of diamond drilling.

Mill Operations—1935

The Idaho Maryland mill operated satisfactorily throughout the year, averaging 225 tons of ore daily. In addition, 75 tons of stored mill tailings were processed daily through the flotation plant.

The stamp mill was prepared to process custom ore from other mines. A ramp was constructed on the south side of the building to allow trucks to dump ore into the battery feed bin. Between 55 to 60 tons of ore were trucked daily from the Bullion mine for milling.

The Idaho Maryland ball mill and flotation plant were so successful that the New Brunswick mill was rebuilt to become an exact duplicate. An 8-foot by 6-foot Marcy ball mill, Dorr rake classifier, Fagergren flotation cells, Deister concentrating tables, riffles, conveyor belts and various pumps were installed. An amalgamating room was constructed on the lower floor. The existing 5-foot by 7-foot ball mill was used as a regrind mill.

The milling capacity was increased to 350 tons per day, with the 20 stamps on standby in case of need. The newly revamped mill began operation on September 27, 1935. The concrete foundations, equipment installations and building enlargement were accomplished by company carpenters and mechanics, under the direction of Walter Merrigan, construction foreman.

Production—1935

During 1935 there were 69,515 ounces of gold produced from 159,091 tons of ore from the two mines. Of this tonnage, 83,137 tons were hoisted through the Idaho Maryland shaft and 75,954 tons through the New Brunswick shaft.

The gross recovered value of the Idaho Maryland-Brunswick ore was $2,452,179, with an average value of $15.41 per ton. The average tailing loss was 72 cents, making the gross value of ore $16.13 per ton.

The total gross income from the sales of bullion and concentrates amounted to $2,558,234, or $16.02 per ton. The total operating expenditure, including development, compensation insurance, taxes, overhead expense, and alterations and repairs to plant equipment was $1,390,015, or $8.73 per ton. The profit from mining operations before depletion, depreciation and federal taxes was $1,168,229, or $7.43 per ton. The total value of custom ore purchased in 1935 was $219,314.

Bullion Mine—1935

The Idaho Maryland Mines Corp. entered into an agreement with the Grass Valley Bullion Mines Co. involving a controlling stock ownership option. The Bullion company lacked sufficient funds to develop the mine, so the corporation agreed to advance funds for specific expenditures. These advances were secured by a deed of trust.

Mining on the 300-foot level was completed and very little ore was left on the 800-foot level. The 1500-foot level north drift had advanced 1,200 feet and a winze had been started. Due to boundary restraints there was only one way to go, and that was down. There was an average of 65 tons of ore trucked daily to the Idaho Maryland stamp mill.

Forbestown—1935

In the early part of 1935 the Idaho Maryland Mines Corp. entered into a new venture involving the lease purchase options of several well-known gold producers near Forbestown in Butte County. The mines and claims were consolidated into one unit of approximately 3,000 acres of contiguous mining claims. These included the Midas and Gold Bank mines, located on the south side of the South Fork of the Feather River. It also included the Carlisle and Denver mines, as well as the Garibaldi and Gold Reef claims, located on the north side of the river.

The veins of this area were greater in width than those in the Grass Valley district, and provided the potential for a large tonnage of low-grade ore. Professor Clifford Tolman, head geologist for the corporation, had been deeply interested in the Forbestown properties, and felt that they would be profitable to mine.

In September 1935 Idaho Maryland engineer Jack E. Ehrhorn was sent to the Forbestown operation as general superintendent. Twelve other men were sent along to begin rehabilitation of the properties. Living quarters, kitchen facilities and storerooms were the first to be built.

General—1936

During 1936 the corporation entered into a lease-purchase agreement with Errol MacBoyle for the operation and maintenance of the Loma Rica ranch and other properties, thereby relieving the corporation from all expense and liability for their operation. Mineral rights under the ranches were retained by the corporation. One of the considerations of this transaction was that the company would have use of MacBoyle's plane for transportation of bullion, company officials, and injured employees in the event of emergencies.

The John H. Hansen ranch property, located between the Idaho Maryland mine surface workings and East Bennett road, was acquired for the mill expansion of a new proposed cyanide-leaching plant.

The capacity of the tailings-impounding lake was increased by adding ten feet of waste rock to the height of the dam.

A suggestion contest on subjects of safety and efficiency was established for employees of the Idaho Maryland, Brunswick, Bullion and Forbestown mines. The contest was inaugurated April 1 and continued to December 30. Substantial awards were handed out to winners in various categories at the end of each three-month period.

The men showed continued interest in the contest, with over 70 miners entering suggestions, some more than once. Many ideas were accompanied by diagrams, sketches and drawings. Assistant general manager Bert Crase presented the awards. The judges were MacBoyle, Crase, Edwin Oliver and Harry Field.

In December Bert Crase addressed a letter to all employees explaining a profit-sharing plan adopted by the board of directors to reward employees by letting them share in the company profits.

It was explained that each time the corporation declared a stock dividend, the directors would set aside an amount equal to the dividend on 150,000 shares of stock. This amount would then be placed in a fund for distribution to the employees on December 15 of each year. They would be paid in units of four, five or six, based upon length of service.

Idaho Maryland Underground—1936

The Round Hole shaft was completed in March 1936. It terminated adjacent to the Mitchell crosscut on the 1000-foot level. The shaft was 1,125 feet deep, with stations cut on the 600-, 850- and 1000-foot levels. The shaft included a 40-foot sump below the 1000-foot level. This type of shaft was the first of its kind in the world.

The company purchased a large bus, and the men working in the far eastern part of the mine, above the 1500-foot level, were bussed from the dry building up to the Round Hole shaft. They were lowered down this shaft in a double-decked cage to their respective levels. At the end of the

The Idaho No. 2 "Round Hole" shaft.

Round Hole station on the 1000-foot level, adjacent to the Mitchell crosscut.

shift these men were transported to the 1000-foot level shaft station by man-train and hoisted to the surface through the Idaho shaft.

The completion of the Round Hole shaft provided a much needed second exit, better ventilation, enhanced movement of timber and supplies, and allowed miners to arrive at their work places sooner. The new shaft provided natural air circulation to the whole eastern portion of the mine through the Round Hole shaft, then down through the mine workings to the 2000-foot level. The air then made its way to the Idaho shaft, where it was exhausted to the surface.

The No. 13 crosscut on the 1000-foot level was started in January 1934. It was run parallel to the No. 3 vein and in firm ground 75 feet in the footwall. The purpose of the development was to explore the eastern continuation of the No. 3 vein by crosscutting back to the vein. This was to avoid expensive timbering and maintenance costs in heavy ground adjacent to the faults along the No. 3 vein.

A separate crosscut, driven back to the No. 3 vein from the No. 13 crosscut, proved the No. 3 vein to be too narrow for mining. The decision was made to continue driving No. 13 crosscut and turning it in a southerly direction, toward the New Brunswick mine. The crosscut was driven over 2,700 feet before being stopped in 1936

Several trunk surveys were run during 1936. An elevation tie between the No. 13 crosscut on the 1000-foot level of the Idaho Maryland and the No. 30 drift on the 1200-foot level of the New Brunswick mine was run to permit possible connection between the two mines. New traverses were carried from the surface down the Round Hole shaft to the 1000-foot level and down the 201 winze and No. 3 raise to the 2000-foot level. Based upon these surveys, full confidence was gained in the correlation of the surface and underground positions of both mines.

A new winze was started on the 2000-foot level, 300 feet southeast of the No. 45 winze. The winze followed the Morehouse vein on an incline of about 40°. This winze was driven to explore the intersection of the Morehouse and Idaho vein systems.

Brunswick Underground—1936

No. 1 winze was started on the No. 1 vein on the 1300-foot level with the intent of opening up the 1450- and 1600-foot levels without interfering with the shaft-sinking operations.

The No. 1 crosscut on the 1300-foot level cut the No. 5 vein and No. 16 vein, where good values were found on the No. 16 vein. The No.

4 crosscut opened up the No. 32, No. 14 and No. 6 veins. Drifts were run on each of these veins. Drifting continued on nearly all veins on the other levels, with raises run to block out ground for stoping. Particular care was directed to running raises to maintain good air circulation from one level to another.

Drifting, crosscutting and diamond drilling discovered 18 distinct veins thus far in the Brunswick mine. Each proved to be of lower grade and narrower than the Idaho Maryland vein systems.

The physical condition of the Brunswick underground workings was greatly improved through natural air circulation. Air doors were installed at all shaft stations, forcing fresh air down the shaft to the 1300-foot level. The air then rose up through raises to all levels before exhausting to the surface through the Old Brunswick shaft.

During the year, 34,460 linear feet of development work was done in the two mines. From this development work and recapture of old workings 130,113 tons of ore and 44,419 tons of waste rock were extracted. There were 13,458 feet of diamond drilling done.

Mill Operations—1936

A new 350-ton-capacity leaching plant was completed in June for treating the sandy portion of the Idaho Maryland mill tailings and the slime portion of the flotation tailings. This process increased the gold

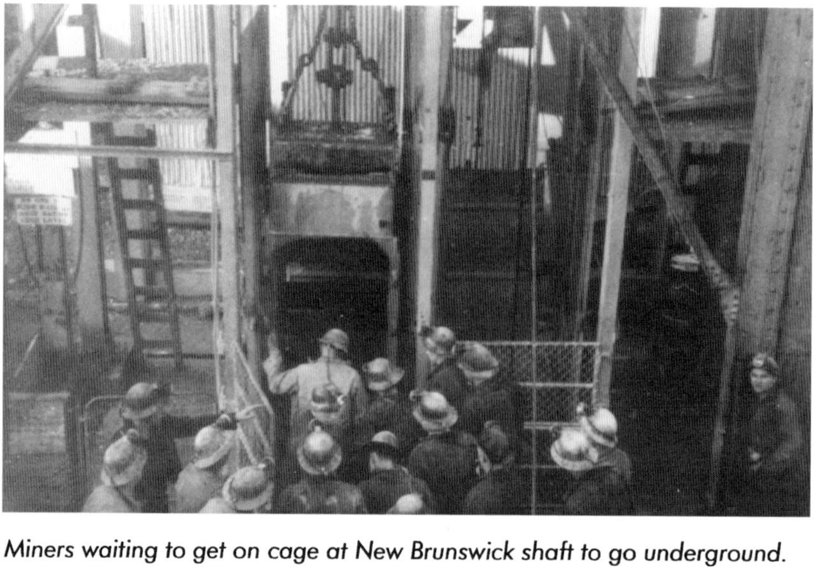

Miners waiting to get on cage at New Brunswick shaft to go underground.

Cage loaded, with both decks full, at the New Brunswick shaft.

extraction by several percent. The new plant acted as a safety valve whenever there was a sudden increase in "high-grade," or a sudden change in character of ore fed to the mill.

The leaching plant was constructed with a semi-gravity flow for treating mill tailings through various agencies of classification, sand tanks, cyanide solutions and precipitation. It consisted of a 72-ton Dorr classifier, six large redwood sand tanks, several hexagonal vats and two steel cyanide tanks. The tailings sands from this operation were pumped to the tailings pond.

The production of the New Brunswick mill jumped from the usual 350- to 400-ton daily output to close to 500 tons per day. This was done by activating the idle 20 stamps for use as a primary crushing unit for the ball mill. The total tonnage of ore treated by all three company mills was just short of 1,000 tons per day.

Production—1936

During 1936 there were 293,975 tons of ore produced from the Grass Valley operation. Of this tonnage 86,753 tons were hoisted through the Idaho shaft, and 207,222 tons hoisted through the New Brunswick shaft. Of this total, 130,113 tons were derived from stoping operations. In addition there were 24,360 tons of old mill tailings reclaimed, 14,504 tons of custom ore treated, and 150 tons of Idaho dump rock milled. A total of 332,989 tons was treated in the three mills.

The gross recovered value of 293,975 tons of Idaho Maryland-Brunswick ore was $3,308,450, with an average value of $11.25 per ton. The average tailing loss amounted to 59 cents. The gross value of ore was $11.84. The total recovered value from the old tailings reclaimed was $51,968, at an average of $2.13 per ton, which, with a tailing loss of 50 cents per ton brought the gross value of the old tailings to $2.63. The total value of custom ore purchased was $144,034.

Total gross income from sales of bullion and concentrates amounted to $3,401,106. The total operating expenditure, including development, compensation insurance, taxes, overhead expense and plant equipment, was $1,864,169. The profit from mining operations before depletion, depreciation and federal taxes was $1,536,841.

Dividends totaling $701,524, equivalent to 40 cents per share, were paid during 1936. The new undistributed profit tax, which became effective during the year, literally forced the board of directors to declare larger dividends than the cash position warranted. The result was that the corporation had to borrow heavily to meet the last dividend payments.

Bullion Mine—1936

The Idaho Maryland Mines Corp. continued to hold an option on a controlling stock ownership in the Grass Valley Bullion Mines Co. Necessary advances, secured by a deed of trust on the Bullion mine properties, were continued. As of December 31, 1936, the total advances amounted to $548,799. Under the option the advances could translate into stock ownership. Operation of the mine continued and was carried out by Idaho Maryland staff members on a time-compensation basis.

A suit was brought by the Empire Star Mines Co., Ltd., against the Grass Valley Bullion Mines Co. and the Idaho Maryland Mines Co. (a predecessor of the Idaho Maryland Mines Corporation). The suit alleged trespass on the Galena vein by activity of the Bullion mine. The Grass Valley Bullion Mines Co. claimed legal rights to the vein and its extralateral rights. By stipulation, the Bullion mine was restrained from mining this vein until the suit was settled by the court.

Forbestown—1936

The initial work at Forbestown was confined to rehabilitating and equipping the well-known Gold Bank and Midas mines, early-day producers in the Butte County district. The two lie about four miles apart, at opposite ends of the district.

Since the first of the year the company had opened work at the Gold Reef and Garibaldi claims on the north bank of the South Fork of the Feather River. The company added considerable acreage in the canyon area by purchase from the Central Pacific Railroad.

During the year, construction was started on a new road connecting the Gold Bank mine with the Midas mine, where truck haulage to the mill was necessary.

The new mill included a Marcy ball mill, classifier, and flotation cell units—almost a duplicate of the Idaho Maryland and Brunswick mills. Construction also included a shop building, crusher building, and a bunkhouse to accommodate 30 men. All building construction was done under the direction of Walter Merrigan.

The mill began operations during the first week of November 1936. Ore from the Midas and Denver mines was readily accessible, so their ore was treated first.

Aerial view of Idaho Maryland facilities, showing cyanide leaching plant on far side of mill building.

CHAPTER 11: LABOR UNREST

The profit-sharing plan offered by the company was inaugurated on January 1, 1937. Two months later contract negotiations began between the mine owners and the Mine Workers Protective League. The League asked for a $1.00 per day wage increase. The Idaho Maryland Mines Corp. offered to pay 50 cents more if they could drop the profit-sharing plan. The employees voted to accept the wage increase in preference to sharing the profits.

Labor unrest began at about the same time as the contract renewal was to take place between miner owners and the League. Jack Shupert, local organizer for the CIO, had begun recruiting miners in the Grass Valley-Nevada City area early in 1937. Many new members were signed up in each of the local mines.

Twin Cities Local 283, International Union of Mine, Mill and Smelter Workers, a CIO affiliate, was organized on March 1, 1937, with 125 new members signed up at their first meeting. The new union's goal was to replace the League as sole agency for the purpose of collective bargaining for wages and working conditions.

The Mine Workers Protective League received a welcome surprise when a group of more than 200 local citizens pledged to support the League as the recognized labor organization. The citizens group, headed by Loyle Freeman, requested mine owners to renew the labor contract with the League.

The company closed the Old Brunswick mine on April 5. General manager Bert Crase stated that ore in the Old Brunswick was marginal and said the closing was for economic reasons. He explained that it would cost less to hoist rock through the New Brunswick shaft, as it was closer to milling operations.

Following the closure, Twin Cities Local 283 filed charges against Idaho Maryland Mines Corp. with the National Labor Relations Board, alleging discrimination against its members. The union said the closing constituted a "lockout" of approximately 80 union workers. The CIO union had 80 members working at the Old Brunswick at the time the mine closed, seven of whom were officers. By then, each of the major mines of the area had signed binding contracts with the Mine Workers Protective League.

National Labor Relations Board—1937

On June 22, 1937, National Labor Relations Board examiner Patrick H. McNally began hearings regarding the charges in the Veterans Memorial building in Grass Valley. NLRB attorney Bertram Edises appeared as counsel for the government, and Judge Edgar Zook for the corporation.

Judge Zook objected to the proceedings before the first witness was called, on the ground that the corporation was not engaged in interstate commerce. He explained that the gold and silver was shipped directly to the U.S. Mint in San Francisco, as required by the National Gold Reserve Act. He also contended that the NLRB had no jurisdiction over the Idaho Maryland Mines Corporation.

Bertram Edises alleged, in answer to the objection, that some of the gold was forwarded from San Francisco to the Denver Mint, thus the corporation became a part of interstate commerce. He also advanced the theory that equipment used by the mines was shipped from other states, and therefore was a part of interstate commerce. The examiner overruled Zook's objection, and began six days of hearings, including testimony from more than forty union and management witnesses.

An interesting aspect of the hearing dealt with high-grading, which was explored extensively. Several men expressed strong disapproval of the "double dry" system that the mine used to minimize high-grading. It was stated that miners were required to remove their work clothes in one room and walk naked to the shower room, then walk naked to another room where their street clothes were hung. It was not deemed unreasonable to have rules to prevent high-grading, but miners felt it was undignified to compel men to strip, as they had to do at the mine. This system was practiced at nearly all of the mines.

Deputy state mineralogist John F. Bongard, mine investigator for the U.S. Bureau of Mines, testified that he and U.S. Secret Service agent Charles B. Rick had been investigating illegal gold transactions throughout northern California. He stated that two men from Nevada County had been convicted as part of an illegal gold-buying ring. He indicated that high-grading was widespread, ranging from $400,000 to $500,000 each year. He estimated the yearly amount of high-grading from the Idaho Maryland and Brunswick mines to be $250,000.

General manager Bert Crase testified that his primary reason for closing the Old Brunswick was due to high-grading problems. He explained that sacks of high-grade were often found on the dump, where they had been tossed by miners coming from the shaft to the dry. The drain tunnel was also used to remove gold from the mine.

Following the hearing, trial examiner McNally forwarded the hearing transcript to the National Labor Relations Board for its review and ruling.

Development—1937

The results of exploration and development work during 1937 were very satisfactory. During the year 27,078 feet of new headings were driven, and 25,878 linear feet of diamond drilling was done. This work resulted in proving the lateral and downward extensions of known ore bodies, and partial development of new veins and ore bodies, both in the Idaho Maryland and Brunswick workings.

Development in the Idaho unit was of two types: the gold-bearing quartz veins, and gold-bearing mineralized diabase dikes. There were 69,391 tons of quartz developed and 103,572 tons of the diabase type. Production of quartz ore was 54 percent from the No. 3 vein, 19 percent from the No. 13 vein, 7 percent from No. 2 vein, and 5 percent from No. 5 vein. New ore was developed in these veins in nearly the same ratio to the total developed.

Production of mineralized diabase ore was 58 percent from No. 22 vein, 25 percent from No. 23 vein, and 18 percent from No. 15 vein. Nearly one half of the ore of this type developed during the year was in the No. 22 vein. The ore reserve at the end of the year was 301,356 tons of quartz, and 93,075 tons of mineralized diabase.

Development headings continued on various veins between the

Miners line up to board the Idaho shaft gondola to go underground.

Miners smile for the camera before going down the Idaho shaft in 1937.

2000-foot level and the 600-foot level. The veins numbered 14, 20, and 122 were important new high-grade veins found in 1937.

At the end of the year the No. 2 winze on the 2000-foot level had been sunk 625 feet on the incline, following the Morehouse vein. An ore pocket and a small station were cut at the 2400-foot level and a drift started.

Development in the Brunswick unit continued with great success from both drifting and diamond drilling. The 1450- and 1600-foot levels were started from the No. 1 winze, sunk from the 1300-foot level. Up until this point, levels had been established 100 feet apart. The greater distance of 150 feet between levels was to provide additional backs between levels and reduce the amount of horizontal mining as the mine advanced in depth.

In 1937 ore developed on the No. 16 vein was 40 percent of the total developed, ore on the No. 18 vein was 30 percent, and on the No. 10 vein was 14 percent. The most outstanding development was finding excellent ore on the No. 16 vein and its branches on the 900-, 1100- and 1300-foot levels. A second development of importance was good ore found on the No. 1 vein in the winze between the 1300- and 1600-foot levels.

Mine Fatality—1937

Edward S. Hall was fatally injured on October 14, 1937. He and his partner, Robert H. Walker, were working in a stope on the 1300-foot level of the New Brunswick mine when a slab of rock fell from the hanging wall. Edward Hall was struck on the head and killed instantly. Walker was brushed aside by the falling rock, and he sustained an ankle injury, along with lacerations and contusions to the head and body.

Both men were removed immediately from the pile of rock by miners working nearby. Walker was taken to Jones Hospital by ambulance, where it was determined that his injuries were not serious. Edward Hall's body was taken in charge by coroner L. Robert Jefford.

Production—1937

The combined production of the Idaho Maryland and New Brunswick mills averaged 1,000 tons daily. The Idaho Maryland mill averaged 300 tons, and the New Brunswick averaged 700 tons a day. In addition, 60- to 75-tons of Bullion mine ore was being milled in the stamp mill at the Idaho Maryland.

During the past year 305,107 tons of ore were produced from the Grass Valley operations. Of this tonnage 95,378 tons were hoisted through the Idaho Maryland shaft, while 209,729 tons came from the Brunswick shafts. Of the total production of 305,107, stoping operations contributed 197,834 tons, and 107,273 tons were from development. There were also 13,627 tons of custom ore treated, of which 12,911 tons came from the Bullion mine.

The gross recovered value of 305,107 tons of Idaho Maryland-Brunswick ore was $3,712,812, an average value of $12.17. The average tailing loss was 51 cents. The gross value of ore was $12.68 per ton.

The total recovered value of Grass Valley Bullion mine ore and custom ore and concentrate purchased was $176,786 for treating, which was received as custom revenue in the amount of $36,914.

The total recovered value of company production from its Grass Valley and Forbestown mines, before mint charges, amounted to $3,994,234, with a head value of $4,206,714, resulting in a tailing loss of $212,480, derived from a treatment of 347,392 tons of ore and 53,625 tons of tailings.

The total income from sales of bullion and from custom milling amounted to $4,036,416. Total operating expenditures were $2,392,296. The profit from mining operations before depletion, depreciation and federal taxes was $1,604,811.

Numerous questions had been asked by stockholders regarding ore reserves, continuity of ore bodies, and life of the mine. The latter question was answered by MacBoyle in this manner:

> "As to the probable life of the mine, from 1926 to date, the Idaho Maryland properties have produced approximately $15,000,000. Most of this amount has come from above the 2000-foot level, from extensions of old veins and ore bodies, or from new veins still being found. From a geological standpoint we believe that many new veins and ore bodies will be discovered in the large area of highly mineralized zone now controlled by the company.
>
> "In the past the Idaho Maryland veins and ore bodies have always been larger, more continuous and richer than any other in the district. In view of the fact that other mines in the district are now working at depths of 10,000 feet on the vein and vertical depths of over 4,000 feet, as compared with depths of 2,000 feet

at the Idaho Maryland, there was every reason, geologically and otherwise that the Idaho Maryland-Brunswick properties would be profitably operated for many years to come."

Bullion Mine—1937

The Bullion mine was closed for the first three months of 1937. This was due to the winter storms and heavy snows. All employees were laid off until the mine could reopen, with the exception of those necessary to keep the mine dewatered. The hoist and compressor were second-hand when they were installed in 1933. Both of these were overhauled during the time the mine was closed.

On January 29, 1937, Judge A. F. St. Sure, of the U.S. District Court at San Francisco, rendered his decision in the case of Empire Star Mines Co., Ltd. vs Grass Valley Bullion Mines. He upheld the claim of the Grass Valley Bullion Mines Co. to ownership of the Galena vein and the right to follow it extra-laterally under the properties of the Empire Star. The Empire Star Mines Co. announced they would appeal the decision.

Subsequent to the layoff, mining resumed on the 1500-foot level and in the winze below the level. The 1700-foot level was established with drifting in a northerly direction. The average daily ore production remained at 65 tons.

Forbestown—1937

Ore from the Midas and Denver mines was most readily accessible to the mill, so it was treated first. Unfortunately, the Midas ore body, which sampling had showed to be of fair grade, proved to be disappointing and not profitable to mine. After three months of operation the mill was shut down until the roads could be constructed to bring ore from the Gold Bank and Golden Queen mines.

Harold F. Lynn was transferred to Forbestown on March 15, 1937. He had been superintendent of the Idaho Maryland mine while Bert Crase was off work with an injury. Lynn held the same title at the Forbestown operations.

The mill reopened in May when road conditions eased and ore from the Gold Bank and Golden Queen mines could be transported to the mill. Considerable road construction and repair was required to make the properties accessible. Powerline extensions and installation of equipment also became necessary.

Active development of the ore vein was accelerated because of its better grade of ore. Tonnage through the mill from the Gold Bank and Golden Queen built up to 300 tons per day. Some months showed a profit and some a loss.

By the end of the year production was running 7,000 tons a month, with an outlook of 10,000 tons a month in 1938, depending upon winter conditions. The mill operated on a three-shift, 24-hour basis. The underground force was increased to 110 men.

During 1937 there were 42,285 tons of ore produced and treated at Forbestown, with a head value of $240,567, or $5.69 per ton. This resulted in a recovery of $217,188, or $5.14 per ton. The concentrates were hauled to the Idaho Maryland cyanide plant for treatment.

The net mint settlement from bullion sales amounted to $216,278, while the operating loss, before depreciation and depletion, of $76,918 further increased to $106,176 after allowance for depreciation and depletion of $29,259. Capital expenditures for the year amounted to $46,391.

National Labor Relations Board Ruling—1938

On January 10, 1938, the NLRB announced its decision relative to the 1937 charges brought against the Idaho Maryland Mines Corp. by Twin Cities Local 283. The board ruled in favor of the union, and supported its claim that the corporation was engaged in interstate commerce and therefore came under the purview of the Wagner Act.

The NLRB ordered the corporation to provide restitution pay for twelve employees who were discharged and later rehired, as well as back pay and reemployment for 61 other employees who had been let go. In addition, the board ruled that the contract between Idaho Maryland Mines Corp. and the Mine Workers Protective League was voided, although it recognized the League as a bonafide bargaining organization.

Under the Wagner Act, this decision could be enforced only by order of a federal court, so the NLRB filed a petition with the U.S. Circuit Court of Appeals for enforcement of its ruling. This action entitled the company to raise all the points of law that it had raised during the original hearings.

On July 19, 1938, the case of the NLRB vs Idaho Maryland Mines Corp. was reversed by the U.S. Ninth District Court of Appeals. The court held that the mine was not engaged in interstate commerce, hence it did not operate under the provisions of the Wagner Act, and the NLRB lacked jurisdiction. The decision was not appealed to the Supreme Court and so the case came to a close.

The Idaho Maryland Mines Corp. signed a new contract for the Mine Workers Protective League after the time limit for the appeal expired. This landmark case was very important to all gold mining companies. Had the ruling not been overturned it would most likely have adversely affected all gold mining companies in the U.S.

Surface Damage—1938

The unusually harsh winter caused considerable damage at the New Brunswick mine. On March 9 strong winds uplifted and partially flattened three of the long garages for employees. The garages were completely open on one side, allowing the wind to enter the structures and lift them off the ground. Several of the miners' cars received damage to windshields and roofs. Several sections of the trestle that supported the tailings flume between the Idaho Maryland mill and the tailings lake were blown down by February winds.

During 1938 the corporation paid off the mortgage on the only piece of property on which there was any encumbrance. Several small surface parcels were purchased, in which the mineral rights were already owned. The surface properties of the company at this point comprised a solid block of 2,180 acres, or about 3.3 square miles.

Underground Operations—1938

The dolly compartment in the Idaho shaft was opened from the surface to the 1000-foot level. The shaft crew rearranged the air and water lines, and installed heavy rails for the gondola.

A single-drum hoist, previously used during construction of the Hoover Dam, was purchased and installed in the hoist room. An 84-inch-diameter sheave wheel was installed atop the headframe to carry the one-inch cable for the gondola. The dolly hoist was used for transporting men, equipment and supplies. This permitted maximum time for hoisting ore by the main hoist.

For several years the ultimate objective had been to connect the Idaho Maryland and New Brunswick mines together underground. Finally, in 1938 development funding became available and the 8,000-foot-long project was started. Work began by drifting east on the No. 1 vein near the No. 30 winze on the Idaho Maryland 2000-foot level. Because of the 230-foot elevation difference at the collars of both shafts, the connection was engineered to occur near the New Brunswick shaft on the 2300-foot level.

Idaho Maryland mill in 1937.

New Brunswick day shift crew, June 10, 1938.

The reasons for connecting the two mines together underground were to provide an emergency exit from the lower levels of both mines, to provide natural ventilation for both mines, and to have the ability to service both mines from either shaft. The long drive would also explore the downward projection of several Brunswick veins.

Exploration and development results had been highly satisfactory in both mines. During the year 32,817 feet of new headings were driven, and 25,394 feet of diamond drilling was done. This work developed some new vein structures already known. The important No. 16 vein developed from the New Brunswick shaft was open on seven levels, from the 580-foot level to the 1600-foot level.

Routine sampling in the Brunswick mine continued during the year. The sampling helped in the estimation of grades of ore from the many points of extraction. A thorough digest of information was made in December. It was found that the sample assays must be raised 1¾ times their reported values to balance the mill recovery. This unusual figure was occasioned by the presence of slugs of gold scattered throughout the quartz.

Sampling in the Idaho Maryland was minimal. Very few veins were sampled. Mining was controlled by whether a vein was ribboned with mineral or not, and by the location of the vein within the overall pattern of the ore shoot. Actually, almost all of the quartz was mined and sent to the mill.

Mine Fatality

Harold H. Hogg sustained a major head injury in an accident on the 1100-foot level in the Idaho Maryland mine. The accident occurred on October 13, 1938, while he was drilling a round in a drift. A large rock suddenly fell from the hanging wall, striking Hogg on the head.

He was administered first-aid and placed in a stokes stretcher, then brought to the surface through the Round Hole shaft. He was transported by ambulance to the Jones Hospital, where he was treated by Dr. Harry Benteen and diagnosed as having critical basal skull fractures. Harold Hogg died in the hospital at 5:10 p.m. on October 14, 1938, without regaining consciousness.

Production—1938

Ore production was greater than in any other year in the mine's history. Larger and better ore reserves were developed, and earnings and dividends reached a new high of sixty cents a share.

During the year 331,406 tons of ore were produced by the Grass Valley operations. Of this total, 113,259 tons were hoisted through the Idaho shaft and 218,147 tons through the two Brunswick shafts. Of the year's tonnage, 231,501 tons were from stoping operations, and 99,905 were from development. There were 20,576 tons of custom ore from the Grass Valley Bullion property, and 70,446 tons from the Forbestown operation. The gross custom revenue was $47,987.

The combined value of the Grass Valley and Forbestown production was $4,616,647, with a tailing loss of $170,827. The tailing loss was 41 cents per ton. The recovered value of company production before mint charges was $4,148,306 from Grass Valley operations, and $297,518 from Forbestown, for a total of $4,445,820. The recovered value of Idaho Maryland-Brunswick ore was $12.46 per ton, and $4.22 for Forbestown ore.

The total income from the sale of gold bullion, including custom milling, was $4,555,374. The total expenses, including development, compensation insurance, property insurance, social security insurance, taxes, mine management overhead, repairs to the plant and equipment, were $2,845,681. The profit on mine operations before depletion, depreciation and federal taxes, after absorbing the operational loss on Forbestown for the year, was $1,665,163.

Bullion Mine—1938

Mining in the Bullion mine took an upswing during 1938, when good ore was found on the 2200-foot level. Small particles of gold were present quite often. The mine was producing an average of 70 tons of ore a day. Howard Dennis was the mine superintendent, and Phil Folck was mine foreman.

The corporation's relationship with the Grass Valley Bullion Mines remained the same, advances being secured by deed of trust on properties, certain blocks of stock, and other guarantees. Advances during the year totaled approximately $72,000, largely to pay maturing property payments. Those payments remaining amounted to about $63,000, payable within two years. No advances were necessary during the last three years, because the mine was on a self-sustaining basis.

The Empire Star Mines Co. appealed the ruling of Judge A. St. Sure in favor of the Grass Valley Bullion Mines Co. relative to ownership of the Galena vein and its extra-lateral rights. The U.S. District Court of Appeals reversed St. Sure's decision and ruled in favor of Empire Star

Mines. The judgment amount for payment for ore mined from the Galena vein was to be forthcoming.

Forbestown—1938

Fred Denton, foreman of the Brunswick mine, and Charles Henwood, shift boss at the Idaho Maryland mine, were appointed superintendent and foreman, respectively, of the Forbestown operations. These men relieved Jack Ehrhorn and Harold Lynn, who had held those positions since the Forbestown operations began.

When the Forbestown group of mines was taken over by the corporation in 1935 it was known that the veins were low in grade. It was felt that a profit could be made if mined in a large production manner. After mining began it was found that the grade of ore was considerably lower than the original test samples revealed.

The Forbestown operations proved to be unsatisfactory to continue mining, and in December 1938 the project was abandoned. All leases and options were surrendered back to their owners. The resultant loss was $213,046. Many of the miners were transferred to the Grass Valley mines. The mining equipment was dismantled and trucked to the Idaho Maryland or Brunswick mines for future use.

Surface Operations—1939

Carbide lights worn by miners were replaced during 1939 by battery-operated Edison cap-lamps. The new lamps provided greater illumination and were well received by the miners. A lamp room was set up at both mines to dispense and recharge the lamps after each use. The men were assigned a lamp by number and used the same one on each shift.

In the event that a lamp was not turned in at the end of the shift, it was quickly determined by the lamp number who the person was that was missing. The safety engineer and shift boss or foreman would go underground to the person's workplace to determine his absence.

For decades the drill bits had been formed directly on the drill steel. The miner needed twenty or more drill steels of various lengths to drill his round. This meant considerable handling of steel, both underground and on the surface.

During 1939 the company changed to using Ingersoll detachable jackbits. The drill steels were threaded on the end, and the jackbits were screwed onto the steel. The bits were easily removed when dull, then placed on a small rack and sent to the surface for sharpening. The same drill steels were used over and over. Therefore, the miner needed only one set of steel to drill his round. Each jackbit could be sharpened four times before it was discarded.

The jackbits were annealed in a furnace and then sharpened on a hot mill. They were reheated to 1,800 degrees in a Lindberg rotary electric furnace. From there the bits were tempered by quenching momentarily in lukewarm water before being immersed in a warm fish-oil solution. The jackbits were sized, strung on racks and sent underground for reuse.

Costly curtailment of mining and milling operations had occurred in past years because of heavy snowfalls clogging the Cascade ditch and cutting off water flow to the mine reservoir.

During 1939 the large Loma Rica reservoir was constructed to alleviate the above problem. This 13-million-gallon reservoir had sufficient water storage to meet all water needs, especially in the event of a water outage. A large power shovel was used to dig the reservoir, and mine personnel were used in all phases of construction. The reservoir was constructed at a cost of $30,258.

The reservoir served the Old Brunswick, New Brunswick and Idaho Maryland mines. It also provided water to the Loma Rica Rancho. A rock walkway extended out over the water to a glass-enclosed gazebo. Railings protected the walkway that also encircled the gazebo. The valves controlling the various pipelines were located beneath the gazebo floor.

Underground Development—1939

Active development of known ore bodies, and exploration for new ore continued through 1939, resulting in 42,356 linear feet (8 miles) of new headings driven. This was an increase of 29 percent over the 1938 footage and was in keeping with a 24 percent increase in tonnage of ore mined. In addition, there were 29,255 lineal feet drilled by diamond core-drills.

The major unusual expenditure for the year was for sinking the New Brunswick shaft from the 1500-foot level to the 2400-foot level, and concurrently extending the 2000-foot level of the Idaho Maryland to connect with the vertical shaft at the New Brunswick 2300-foot level.

One half of the total footage of new headings in the Idaho Maryland was driven on or below the 2000-foot level. This reflected the general opinion that new ore would have to come from the downward extension of veins numbered 1, 2, 3, 4 and 5.

To explore these extensions it was necessary to dewater and retimber the No. 30 winze. It had been sunk 850 feet on the incline to the 2400-foot level, just before the Metals Exploration Co. closed the mine in 1925. Very little mining had taken place at that time, and the winze had been allowed to fill with water.

The veins explored in the No. 30 winze area were disappointing because of the low-grade quartz encountered. Exploration on the 2400-foot level found the downward extension of the faulted No. 5 ore body, and gave assurance of extensions of ore below the 2000-foot level.

A decision was made to extend the winze another 500 feet. The three compartments were cleaned out, retimbered and readied for sinking operations. The first round occurred on November 15, 1939.

The long drift connecting the two mines together on the 2000-foot level was nearing completion by the end of the year. The No. 1 vein, which the drift followed, did not show any evidence of ore at any point, and was quite disappointing. Crosscuts were driven from the drift to intersect the downward extension of the six Brunswick veins, which also were very disappointing.

The 3,000-foot long Mitchell crosscut was stopped after only five narrow low-grade veins were cut. A raise was run up to the 920-foot intermediate level on one of the veins, and 200 feet of drift run on that level. A small stope between the 1000- and 920-foot levels produced only low-grade ore. The Mitchell crosscut, driven from the Idaho Maryland 1000-foot level, passed directly underneath one of the horse barns of Errol MacBoyle's Loma Rica Rancho.

Shaft sinking in the New Brunswick shaft continued without interfering with general mining activities. Station cutouts for future levels were made at 140-foot intervals as the shaft-sinking operation advanced.

By November the three-compartment shaft had reached the 2300-foot level, where underground connection between the two mines was to take place. The shaft crew continued sinking, while other miners enlarged the 2300-foot level station and prepared the ore and waste pockets.

In addition to the shaft sinking, drifts on nearly all levels in the Brunswick mine were extended during the year. Crosscuts were extended into the footwall on the 1450- and 1600-foot levels to develop the downward extensions of the veins numbered 28, 30 and 31.

Mine Fatality—1939

On November 30, 1939, Jim Patterson and Edward Keleher were replacing the lip on the skip-loading chute located below the 1500-foot

Motormen loading ore cars from ore pocket at No. 30 winze on the 2000-foot level of the Idaho Maryland mine.

level station in the Idaho shaft. They had talked with the hoistman and obtained clearance to work in the west shaft compartment, and would be using the gondola to work from.

Two mechanics were engaged in making sheave-wheel adjustments at the top of the headframe at the same time that Patterson and Keleher were working in the shaft. When the two mechanics completed their work on the east compartment sheave wheel and switched over to the west compartment sheave wheel, they signaled to rotate the sheave wheel in a counter-clockwise direction, which caused the gondola in the west compartment to move quickly upward. Keleher was thrown from the gondola and down into the steep incline shaft.

Patterson was thrown into the adjoining compartment, but was able to cling to a timber, thus preventing his own fall down the shaft. Patterson was able to climb to the shaft station and telephone the hoistman to report the incident and obtain help. Edward Keleher was found approximately 200 feet below where he had been working. He was brought to the surface at once by fellow workers, and pronounced dead by Dr. Harry Benteen. He had died instantly from the fall.

The two mechanics said they were under the impression that they had clearance to work on both the east and west sheave wheels, which would have affected both shaft compartments. There was an obvious misunderstanding between the mechanics and the hoist engineer. This tragedy could have been prevented had the two work assignments been scheduled at separate times.

Production—1939

The average daily tonnage of ore milled during 1939 was 1,435 tons. Of this total, the Brunswick was the greatest contributor, with a daily estimate of about 1,000 tons. Of these, 675 tons were milled at the New Brunswick mill, and the remainder trucked to the Idaho Maryland mill. The Idaho Maryland production was 360 tons. The Bullion mine maintained a constant average of about 75 tons per day, which was milled in the stamp mill.

Tonnage of ore milled was by far the largest in the history of the company—410,411 tons, or nearly 25 percent more than in 1938. Fairly large quantities of relatively low-grade ore and more ore from both mines were milled. As a consequence of dilution from these two causes, the average grade of ore milled during the year was lower than in 1938, but the drop in grade was largely offset in increased tonnage milled. In spite of the large increase of tonnage hoisted from the Brunswick shafts, the average grade of ore from that mine was 26 cents higher that in 1938.

The Idaho Maryland mine contributed 119,611 tons, and the Brunswick mine 290,800 tons. Thirty percent of the 1939 ore production came from development headings, the remainder came from stoping. The stamp mill treated 23,813 tons of ore from the Bullion mine and 288 tons of concentrates, from which $36,550 was received.

There were 115,001 ounces of gold recovered. This represented an average of $10.19 per ton, with a tailing loss of 38 cents per ton. The total income from the sale of bullion, including custom milling, was $4,153,597. The total operating cost was $2,694,619. Profit before deductions for taxes, depreciation and depletion was $1,454,420, with net earnings totaling $955,172. This was equivalent to 53 cents per share of 1,780,781 shares outstanding.

According to government reports the Idaho Maryland Mines Corp. had the largest gold production of any quartz mine in California during 1939. The corporation was the second largest gold mine in the United

States, being exceeded only by the Homestake Mining Co. in South Dakota.

Harmonious employer-employee relations existed during the year. In December the Mine Workers Protective League agreed to accept the same collective bargaining agreement for the coming year, ending December 31, 1940. Labor turnover was less than two percent a month during 1939. The number of employees increased from 781 to 831. A Christmas bonus was paid in December. Payments ranged from $10 to $50, based on the employee's time with the company.

Bullion Mine—1939

The relationship with Grass Valley Bullion Mines Co. remained the same, except to the favorable extent that throughout the year it operated on a profitable basis, and paid property payments out of its earnings. In the past the Idaho Maryland Mines Corp. had loaned money for this purpose.

In March 1939 the Idaho Maryland Mines Corp. purchased the $61,569.82 judgment against the Grass Valley Bullion Mines Co. in favor of the Empire Star Mines Corp., Ltd. The corporation took assignment of the judgment in order to protect the deed of trust covering its past advancements to Bullion Mines. Payments totaling $28,000 were collected during the year, leaving a balance of $33,569.82 due on the judgment as of December 31, 1939. Mining continued at the Bullion mine, with a daily ore production of approximately 75 tons.

1000-foot head pump at the 1300-foot level of New Brunswick.

CHAPTER 12: MODERNIZING NEW BRUNSWICK MINE

The greater shaft depth and anticipated increase in ore production meant increasing the head-frame size and hoisting capability. Then in May 1940 the company embarked upon a modernization of the New Brunswick surface plant. A 1,000 hp Ottumwa double-drum hoist was purchased from Los Angeles Metropolitan Water District. This hoist had been used during construction of the Parker Dam on the Colorado River.

The hoist had eight-foot-diameter drums with a 60-inch face capable of holding 4,800 feet of 1⅜-inch cable. A 600 hp Nordberg single-drum hoist was also purchased from the same water district for use with the dolly compartment. This drum was eight feet in diameter with a 60-inch face that held 3,800 feet of 1⅜-inch cable.

All of the cement piers, foundations and drum pits were completed and the large hoists set in place during the summer months. Concrete floors were poured and an all-steel building, 56-feet by 60-feet by 15-feet high, was constructed over the hoists. The new hoist room was located north of the three-compartment shaft, whereas the regularly-used hoist was located on the south side.

The two-story dry building was enlarged to accommodate 400 miners comfortably, instead of the earlier crowded limit of 275 men. The new addition provided ample showers and lavatories, and greater heater and boiler space. The two-story structure separated the miners' clothes so that underground "diggers" were on the bottom floor, and street clothes were on the upper floor. The new addition also provided office space and change rooms for the mine superintendent, foreman, shift bosses and engineers.

Two new 425 cu. ft. Ingersoll-Rand air compressors were installed to meet the anticipated increase in production and demand for compressed air. The new compressors were set in close proximity to the two already in use, and the building was enlarged.

The Cahill Construction Co. of San Francisco was awarded a contract in August 1940 to build a new 135-foot-tall steel headframe. Government priority for steel, caused by a sudden increase in prewar military contracts, delayed the start of the construction. National defense needs took precedence over private industry for all available steel. The headframe project was postponed several times before it was decided to delay the work until the following spring.

Advantage was taken during the steel delay to construct concrete footings for the long support girders on the north side of the headframe. One was positioned in the mill building, which required several modifications, and the other was located just east of the mill building.

A new crushing plant was built, containing an Alloy 15" by 38" roller-bearing jaw crusher and a Traylor three-foot gyratory reduction crusher. The ore was fed from the ore bin to the crusher plant by a 30" belt conveyor leading to a Robins vibrating screen, where the undersized rock was separated. A picking-belt located ahead of the crushers was sized for two men to remove the waste rock into a large waste bin below. A water spray and dust collection system was provided for dust control. The new crusher plant began operation in August 1940.

A large substation located just west of the mine was constructed by Pacific Gas and Electric Co. crews to provide additional power required for the surface changes.

Underground—1940

On January 3, 1940, the long drift driven from the Idaho Maryland "holed through" on the New Brunswick 2300-foot level near the shaft. The engineering was so accurate that the connection was essentially perfect.

The connection was of vital importance for the operation of both mines. A second exit was provided for the deeper workings of both mines. Natural air circulation was greatly increased, with fresh air de-

Setting the Ottumwa double-drum and Norberg single-drum hoists in 1940.

Elmer Olson, hoistman at the New Brunswick, inspects the powerful 1000 horsepower Ottumwa double-drum hoist.

600 horsepower Nordberg single-drum dolly hoist.

Ingersoll Rand compressor.

scending the New Brunswick shaft, then exhausting through the Idaho shaft. Ore haulage could then be made to either shaft to balance hoisting or milling, as desired.

Following the connection of the two mines, the waste rock from the New Brunswick shaft-sinking was trammed to the Idaho shaft. This permitted greater hoisting time in the New Brunswick shaft. Mine water from below the 2300-foot level was pumped to that level and allowed to flow in an open ditch to the Idaho shaft, where it was pumped to the surface.

Development work in the Idaho Maryland during 1940 centered mainly below the 2000-foot level in the No. 30 winze area. The winze was extended downward another 500 feet, and the 2700-foot level established. A station with ore and waste pockets was put in place. The No. 3 vein on the 2070-foot level was explored, but the vein was barren. Drifts were run on the No. 5 vein on the 2140- and 2400-foot levels. In each instance the vein followed the "L" fault. The quartz was broken up along the fault zone, and the assays were very disappointing.

The No. 23 vein was cut on the 2000-foot level by the No. 20 crosscut. This new vein was rather narrow, but demonstrated great potential, as it showed free gold persistently, and the quartz yielded excellent assays.

Development work in the Brunswick mine during 1940 centered around the No. 16 vein between the 580- and 1600-foot levels. Considerable drifting occurred on the No. 28 and No. 30 veins between the 1200- and 1600-foot levels. Exploration began on the 2300-foot level, and the projected No. 1, No. 620 and No. 632 veins were drifted upon.

In April 1940 the shaft-sinking hoist was moved from the 1600-foot level to the 2300-foot level station, where it was installed for further shaft-sinking operations. The shaft had reached a depth of over 2,800 feet by the end of the year, and the objective was to continued sinking to 5,000 feet.

Christmas Bonus—1940

For the second consecutive year, 950 employees received a Christmas bonus in their paychecks. The bonus payment was based upon time with the company—$10 to men working five months, $20 to those employed a year or longer, $30 to men employed over three years, and $50 to those who had worked five years or more.

A concert by miners singing traditional Cornish carols occurred on

The hoisting compartments of the Idaho shaft at the 2000-foot level station. Dolly compartment is at the far right.

Ed Farley operating a hoist in 1939.

A "Christmas tree" (for sinking) carries four or five jackhammer drills (with air and water hoses attached), plus miscellaneous extras. It is shown suspended at the New Brunswick 2300-foot level shaft station in 1940.

Christmas Eve 1940 at the 2000-foot level station in the Idaho Maryland mine. Approximately forty members from the California Cornish Gold Mining Singers and the Grass Valley Carol Choir participated, under the leadership of Harold George.

The program was broadcast live over San Francisco radio station KGO and the National Broadcasting Co. network. The importance of the broadcast can be estimated from the fact that NBC donated the broadcast time. Scores of favorable letters received locally and by many NBC stations showed the success of the Christmas Eve concert.

Arrangements for the broadcast and program were carried out by the Grass Valley Chamber of Commerce, under the direction of secretary Edith Scott and president William L. Goggin.

Production—1940

For the second consecutive year, according to figures reported by the U.S. Bureau of Mines, the Idaho Maryland Mines Corp. was the leading gold producer in California, and the second in the nation. This achievement represented output from the Idaho Maryland, Brunswick and Bullion mines. The Brunswick mine became the largest tonnage producer of any gold mine in the state, with approximately 1,000 tons of ore milled daily.

It is interesting to note that since Errol MacBoyle took over this mine in 1925, over two million tons of ore had been produced, with a value of approximately $28,000,000. This tonnage, and developed ore reserves, were developed as a consequence of successful exploration, conducted under a program through which the owners' capital was subject to a relatively high risk of loss. The development program was conducted under the guidance of a competent staff, and backed by the ardent faith of management.

The total tonnage of ore from the two mines was 406,707 tons. The recovered value of company production, before the mint charges, was $4,549,932, with a head value of $4,732,211, resulting in a tailing loss of $182,720, or 44.3 cents per ton. The average recovery of company ore was $11.19 per ton.

Bullion Mine—1940

The Bullion mine continued its daily production of between 70 to 80 tons of ore per day. Exploration followed the Bullion vein to a depth approaching 2,700 feet. The vein ranged from five to six feet in width. To

The Grass Valley Carol Choir and the California Cornish Gold Mining Singers were heard nationally in a 1940 broadcast from the 2000-foot level.

bring the ore to the surface, two long winzes and the main shaft were traveled, making three separate handlings necessary.

Grass Valley Bullion Mines Co. became heavily indebted to Idaho Maryland Mines Corp. for various services, including underground development work. Repayment from mine production could not keep up with advancements made. During the year the Idaho Maryland Mines Corp. acquired control of the Bullion Mine. Numerous local investors who had purchased shares of Bullion stock sold their shares to the Idaho Maryland Mines Corp. under an offer to purchase that expired on May 1, 1940.

Labor Problems Looming—1941

Entering the year 1941 the Idaho Maryland Mines Corp. was at its peak. The underground development work was well ahead of stoping operations, with good ore reserves in place. The combined number of employees at the Idaho Maryland, Brunswick and Bullion mines was 973. The underground work was expanding, and additional miners were being hired.

Two major projects were well underway. The New Brunswick shaft-sinking operation had reached a depth of 3,000 feet, and the New Brunswick surface modernization project was at its midpoint in completion. The Mine Workers Protective League's wage and hour contract with the local mines had expired on December 31, 1940, and negotiating meetings were underway. There was no demand for a wage increase. The only new request was for a closed-shop agreement. This meant that no one could be employed, or continue employment, unless he became or remained a member of the union.

The mine owners were adamant against granting the closed shop. They were willing to recognize the Mine Workers Protective League as the sole authority to bargain for its members, and agreed to recognize no other union as a bargaining agency for employees.

The League countered by voting to strike all of the mines in 15 days. The deadline date was set for February 7, 1941. The decision to strike also included pulling hoistmen and pumpmen from their jobs. League president Olney Donnelly declared that the men were standing firm on the closed-shop issue, and intended to carry it through.

The Idaho Maryland Mines Corp. exercised its right to hold a vote among its employees to determine whether they desired the Mine Workers Protective League as their bargaining agency. In a special election, held under the nonpartisan auspices of County Clerk R. Noble McCormick, the Mine Workers Protective League won an impressive vote as the choice. As a side issue, and for information purposes only, the preferential vote on the question of a six-day work week or a five-day work week, there was an overwhelming desire for the longer week.

After a two-day negotiating meeting between management and the League, agreement was finally reached. A proposal by the mine owners recognized the Mine Workers Protective League as the exclusive bargaining agency for all workers except supervisory, office and engineering personnel. A closed shop was not included; instead, the new agreement gave the Mine Workers Protective League the right to have its organizers contact the miners during lunch hours, in the dry, or at shaft collars for the purpose of obtaining new members.

The new proposal was placed before the League membership and was accepted by a large majority vote, thus averting the strike. The 18-month contract was signed on February 19, 1941.

Wildcat Strike—1941

Labor troubles began suddenly on Sunday morning, April 27, 1941, when 310 Idaho Maryland Mines Corp. employees held a meeting at the Veterans building in Grass Valley. They convened to discuss the Wage and Hour Administration's order for a forty-hour work week, beginning on May 1.

The meeting was presided over by Homer Owens, a Brunswick mine employee. The group voted to ask management for an eight-hour day, five-day work week, with the same rate of pay received for six days, and time-and-a-half for all hours over 40. This wage increase would amount to $1.24 a day. A walkout was planned for May 1, in the event that their demands were not met.

A negotiating committee composed of Evard Le Du of the Bullion mine, Clifford Sonne and Harold Scribner of the Idaho Maryland mine, and Carl "Louis" Tobiassen and Edward Hamilton of the Brunswick mine, was appointed to meet with general manager Bert Crase to present their demands.

The committee met with Crase, who explained that the company could not negotiate with their group because it had a binding contract with the Mine Workers Protective League.

The request for higher wages and a threatened walkout had its first repercussion when Bert Crase announced that operations at the Bullion mine were being suspended permanently. He stated that the Bullion mine could not operate under a higher wage scale. Fifty men were laid off as a result of the mine closure.

Meanwhile, Olney Donnelly, president of the League, stated that all its members would be expected to report to work, even if a picket line was present.

The Scotia mine was added to the list of mines to be picketed. On May 1 picket lines were set up at the Idaho Maryland, Brunswick, Bullion and Scotia mines. By 7:00 a.m. Idaho Maryland Road was lined with cars as men headed for work—likewise at the Brunswick mine. One surprise element was the presence at the Idaho Maryland entrance of Captain Joe Blake, of the California Highway Patrol, and 14 other patrolmen. Also on the scene were Sheriff Carl J. Tobiassen and several deputies. The strike area at each mine was very orderly and without incident. But, despite president Donnelly's expectation, League members refused to cross the picket lines.

Meanwhile, the independent miners made application through Carpenters Local 1903 for affiliation with the American Federation of Labor (AFL). Charles F. Daley, organizer for the AFL from San Francisco, assisted, and Gold Miners Union 22742 was formed. Edward Hamilton was elected president of Local 22742.

The Mine Workers Protective League and the mine operators reopened negotiations on a proposal submitted by the League. This called for a 48-hour work week, with time-and-a-half for all hours over 40, and time-and-a-half for Sundays and holidays. For the first time since the labor trouble started, Empire Star Mines Co., Ltd. entered into the negotiations. The mine operators approved the proposal, which became effective on May 3.

Mine Workers Protective League president Olney Donnelly requested the U.S. Department of Labor to investigate the strike situation. Omar F. Hopkins, a federal conciliator, was sent to Grass Valley to meet with all parties involved. He held separate meetings with management and both labor groups and found them willing to talk—but not to each other. Hopkins was unsuccessful in bringing about a settlement.

The striking miners dictated the tempo of the labor conditions for over 18 days. The events began to change when 40 miners returned to work at the Idaho Maryland mine. In addition, striking AFL members returned to work at the Scotia mine.

Strikers gather on Idaho Maryland Road on the morning of May 1, 1941. This view looks east, toward the employee garage and the mine buildings.

Later in the day, spectators can be seen on the narrow gauge railroad tracks on right side of this photo that looks westward to Grass Valley.

The 19-day-old strike ended abruptly on May 19, 1941, when over 400 striking AFL members voted almost unanimously to end the strike. The return to work was made with absolutely no concessions on the part of management. All men returned to work under the Mine Workers Protective League contract which had become effective May 3.

The company's payroll had reached over 1,000 employees when the strike occurred, but the world crisis began to have an adverse affect upon the corporation from that time on. War had begun in Europe in 1939 and, although the U.S. had remained officially neutral, it was providing arms to Great Britain and Russia under a lend-lease agreement. In 1940 United States policy shifted from neutrality to "preparedness." The Selective Service Act was enacted to build up the nation's military. Defense plants began building ships, tanks and airplanes, along with a multitude of other war materials. Government priorities were quickly placed on steel, explosives and other products needed for the preparedness effort.

Many men took advantage of their time off during the strike to seek higher paying jobs in shipyards, defense work in Hawaii or in other manufacturing plants. The mine payroll was reduced by at least 60 men within a very short time, most of whom sought work in defense plants, but others entered the armed forces by enlistment or the draft.

Brunswick Headframe—1941

Construction of the 135-foot-high New Brunswick headframe took place during the early summer months of 1941. The new structure was able to be erected over the existing 85-foot headframe, because of its greater width and height. This allowed production to continue with a miniumum of interruption. The Old Brunswick shaft was used to raise and lower men during the changeover from one headframe to the other. Ore was also hoisted through the Old Brunswick shaft and trucked to the Idaho Maryland for milling.

The completion of the New Brunswick headframe, crusher plant, hoists and new compressors, at a cost of $135,000, gave the corporation one of the finest and most efficient mining plants in the U.S.

The 4" by 4" wooden guides in the three compartments of the New Brunswick shaft were replaced by 4" by 8" guides, to compensate for the increased hoisting speed. These were made of straight-grained knot-free Port Orford cedar, and each section measured 37½ feet in length. They were transported to the Brunswick spur track on two railroad flatcars, with one end on one car and the other end on a second car.

The new speed for hoisting men was established at 1,080 feet per minute. The speed for hoisting rock was 1,250 fpm. Each hoist was equipped with a Lilly controlled-braking system that automatically locked the brakes in the event of a power failure, or if the maximum hoisting speed was exceeded.

A limit switch was located 100 feet below the shaft collar in each compartment, which activated an audible alarm to warn the hoistman that the cage was nearing the surface. An override switch was located in each hoisting compartment on the headframe that required the hoistman to physically open the switch to allow the skip to pass by that point. The hoist brakes locked automatically in the event that the override switch was not activated. This was a safety protection to prevent the skip from being pulled up into the sheave wheel and cause major damage.

Communication between the hoistman and the person on a cage was by a bell system. A list of the designated bell signals for each level was posted at each shaft station, along with a set of standard signals established by the California Division of Industrial Safety.

A "call bell" signal system, separate from the shaft signals, was present on each station for use by persons wishing to come to the surface or to go from one level to another. Each station had a telephone that was connected directly to the hoist room, which was used to report an underground emergency.

Unfortunate Fatality—1941

On September 11, 1941, William F. Weiss was engaged in barring down loose rock in a stope on the 900-foot level of the Brunswick mine. He accidentally fell into a chute, causing loose rock to fall around and partially cover him. His partner, Frank Giovanetti, immediately summoned help from miners working close by. They had begun pulling rock away from Weiss when additional loose rock slid down the footwall and engulfed the trapped man.

William Weiss died of suffocation before the men were able to free him from the chute. His body was brought to the surface and taken in charge by Coroner Andy M. Holmes. A rock had struck Frank Giovanetti's shoulder when the second rush of rock occurred, and he was taken to Jones Hospital by the Hooper-Weaver ambulance. His injury was not serious.

On December 7, 1941, the United States was drawn into World War II when Japan attacked Pearl Harbor. This was the beginning of an accelerated loss of manpower from the mines to the armed services and defense industries. Over 200 men had already left the two mines during the previous eight months.

The Grass Valley Carol Choir, under the direction of Harold George, once again sang Cornish carols from the Idaho Maryland 2000-foot level shaft station on Christmas Eve 1941. The program was recorded and then broadcast by radio station KNX. This was the second consecutive year that the choir presented their program from underground. Arrangements were made by William Goggin and the Grass Valley Chamber of Commerce.

Concluding a 10-day legal call for bids for the purchase of the Bullion mine, the Idaho Maryland Mines Corp. was the successful bidder. The purchase was consummated on December 26, 1941. The corporation had operated the Bullion mine for several years previously and already owned a considerable amount of its stock.

A Christmas bonus was paid to employees for the third consecutive year, and for the third year in a row the corporation was the largest lode gold producer in California. Once again, it was second in the nation, outranked only by South Dakota's Homestake mine. A total of 263,769 tons of ore was milled, yielding 113,973 ounces of gold, for an average recovery of $15.20 per ton. The net profit for the year was $1,008,315, with a dividend of 56 cents per share.

Completion of the Modernization Program—1942

The Brunswick mill was placed back in operation during the first week in January 1942, after being closed since the previous May. The mill was completely rebuilt inside during the shutdown, to accommodate one of the large new girders. The mill equipment was rearranged to correspond to the flow sheet. The 20-stamp mill section was removed to make way for the large ore bin foundation.

Construction of the large concrete ore bin within the confines of the new headframe began during April 1942. It was circular in design, and was divided into two sections, the larger portion for ore and the smaller for waste rock. The structure rose 90 feet above ground level, and the interior was lined with heavy steel rails to protect the concrete from wear. A large chute and diversion gate were located between the automatic skip-dumping rails on the headframe and the two bins. This was to direct the ore or waste to the proper storage bin. The large concrete storage bin, often referred to as the "silo" because of its unique shape, was completed in mid-June.

The corporation sought to include strategic or essential minerals in its production, and pursued a diligent search for tungsten in the Union Hill veins within the Brunswick mine, but no profitable deposits were discovered. The corporation also took an option on a quicksilver property near San Jose in the New Almaden district. A fair grade of cinnabar ore was found at the surface, but the deposit was too shallow and lost its commercial value, so the option was dropped.

The exodus of miners and other key personnel increased dramatically following the Pearl Harbor attack. This affected the underground production and the selection of ore to be mined. Government restrictions on the purchase of pipe, rail, drill steel, explosives and mining equipment also had a great impact upon the gold mining industry. The company's low priority-rating was a major concern in obtaining materials.

The management policy always had been to maintain an even grade of ore to be milled at both mines by mixing the lower grade of rock with the better grade of ore. This made it possible to keep tonnage high by including the marginal grade rock while still making a profit. The life of the mine was also extended by this practice. Now, in 1942, the company was being forced to mine only the higher grade ore to offset the decline in production.

The newly discovered No. 23 vein, located on the far eastern end of the Idaho Maryland 2000-foot level, revealed great potential. The vein was narrow, but showed free gold quite consistently. A winze was started and then stopped after about 35-40 feet in depth, because of the shortage of miners. The author operated the hoist during the sinking operation. The area was quite a distance beyond fresh air, and had very little air circulation. The area was warm and had a distinctive odor, thus was referred to as the "Rose Garden."

The New Brunswick shaft-sinking operation was stopped early in the year at a depth of 3,470 feet, far short of the goal of 5,000 feet. Many of the development headings in both mines had to be suspended, and the miners were transferred to work in stopes to keep production flowing.

Government Mine Closure—1942

On October 8, 1942, the War Production Board issued Limitation Order L-208 which closed down all major gold mines. This order required gold mines to stop breaking rock on October 15, and also stop all milling of previously broken rock within 60 days. A minimum number of employees was permitted to remain for pumping water and for essential mine maintenance.

The War Manpower Commission followed with its order forbidding the employment of the jobless miners without government approval. The same order required every man laid off to register with the nearest U.S. Employment Service office for work in the strategic mining industry, which included copper, zinc or molybdenum mining. Latitude was extended to the miner to request placement at a mine of his choice.

Prior to the issuance of L-208, the Idaho Maryland Mines Corp. payroll had been reduced from 800 employees in December 1941 to just 212 on October 8, 1942. The average age of the remaining employees was 47.7 years, and 43 of the underground workers were 50 years or older.

The local economy was greatly affected during the latter part of 1941 and all of 1942, because of the miners and their families who moved away. The closing of the mines made the situation even worse, for it meant additional loss of revenue. This was partially offset by employment becoming available at Camp Beale, near Marysville, and DeWitt General Hospital at Auburn. The Lava Cap mine outside of Nevada City was given a strategic mineral classification by the War Production Board because of the silver it produced along with its gold. The Lava Cap was able to continue working, and in turn was helpful to the local economy.

The Grass Valley Carol Choir, directed by Harold George, sang eight Christmas carols from the Idaho Maryland's 2000-foot level on Christmas Eve, and once again the performance was recorded and broadcast by radio station KNX.

The year 1942 was a hectic year for Idaho Maryland Mines Corp., and that fact was reflected in production records. There was a 700 percent drop in net earnings, from a high of $1,008,315 in 1941 to $147,975 in 1942, with dividends suspended in February. Production tonnage in 1942 was 144,639 tons of ore (against 263,769 in 1941). The average recovery in 1942 was $12.96 per ton, compared with $15.20 in the previous year. On the good side, three new ore bodies were discovered during the year. Idaho Maryland Mines Corp. slipped to fourth place in the state from being the leading gold producer in California during the previous three years.

Errol MacBoyle's Contribution to the Community

By the early 1930s the corporation controlled 2,180 acres of land, not all of which were suited for mining purposes. Local citizens often criticized companies for the unkempt condition of their properties. To remedy this, Errol MacBoyle began purchasing portions of the land for his personal development, thereby releasing the corporation from its responsibility. The mineral rights remained with the mine.

In 1933 MacBoyle began construction on an 1800-foot-long airfield, which was completed in 1934. A large airplane hangar, which fronted the airstrip, was constructed by mine carpenters. Being a strong believer in providing jobs for young people, MacBoyle hired local youths during school vacations to clear brush for the runway and proposed roads, prior to construction.

He then purchased a nine-passenger Vultee V-1A monoplane, which he leased to the company. The primary purpose of the airstrip was for flights to his office in San Francisco, and to fly the gold to the mint. The flights were from Grass Valley to Mills Field at San Francisco, then by automobile into the city. Security was maximized by employment of a full-time radio operator, who kept in constant touch with each flight.

Not a pilot himself, MacBoyle hired two of the best pilots, Latham A. "Slim" Perrett (1934–1937) and James W. Carson (1937–1941). In 1936 the Vultee monoplane was replaced by a Lockheed 10A Electra that featured all-metal construction and a distinctive dual-tail design. A year later the plane was traded back to Lockheed for the newer Model 10E, identical in appearance, but with 600 hp engines. Only twenty of the 10E Electras were built, one of which was flown by Amelia Earhart on her ill-fated around-the-world flight in 1939.

As the war clouds gathered in Europe and the Pacific, the United States began confiscating high-performance airplanes from civilian operators for government use. To avoid this, Errol MacBoyle sold his Electra to Superior Oil Company for the full purchase price of $85,000. The army then took the plane from Superior Oil and it was flown to McClelland Field for shipment overseas. The mine continued its gold flights with a five-passenger Waco plane powered by a single 350 hp engine.

In 1939 and 1940 the California Air National Guard conducted training maneuvers at the airport using Douglas B-18s, which could best be described as the bomber version of the famous DC-3. Various observation planes were also used. The Army Air Corps used the private Loma Rica field as a squadron training site in 1941.

America's entry in World War II brought about the closure of MacBoyle's Loma Rica airport. A ban was placed on all civilian flying within 150 miles of the west coast. The army then dug ditches across the runway to immobilize its use.

Loma Rica Rancho

In 1934 Errol MacBoyle purchased the rundown 452-acre Conti ranch from the Idaho Maryland Mines Corp. The ranch was renamed Loma Rica Rancho (Spanish for "rich earth.")

MacBoyle built two large barns on a knoll overlooking the rest of the ranch. Each barn was 50 feet wide by 150 feet long, with 19 stalls in each. The stalls were 15 feet square, and the walls were lined with 2" x 6" hardwood. There was a door from each stall leading into a long center corridor. Each stall also had a door leading to the outside, in case of fire. The hayloft was floored with 1" x 4" pine flooring.

Each stall had a sunlamp for stimulating growth of the young colts. Each barn had a maternity ward, which made visitors gasp in wonderment. All surgical and medical equipment that could be found in a horse hospital was provided. Johnny Semas, superintendent of the ranch, had sleeping quarters next to the ward to see how the mother and foal were doing.

The 22 paddocks were separated from one another by white wooden fences, and each paddock was planted with Kentucky bluegrass.

MacBoyle purchased a prize string of Percheron horses and began breeding Percherons. These were raised for show, with the exception of those used as work horses around the ranch. His Percherons were exhibited at the California State Fair in Sacramento, and the Los Angeles County Fair in Pomona. Several blue ribbons and awards were won in 1935, 1936 and 1937. He also entered his horses in the 17th District Agricultural Fair, held behind the Veterans Memorial Building in Grass Valley in the mid 1930s. The Percherons pulling wagons on the Hennessy field were a great attraction for the small fair.

MacBoyle auctioned off his Percherons in 1937 and turned to breeding throughbred racehorses. He obtained 24 thoroughbred brood mares and two famous racehorses for stud. These were "Microphone," who won 12 races out of 12 starts, and "Time Supply," a stallion famed from coast to coast. Time Supply was one of the fastest horses of his time. Each new foal was named with the prefix "gold," such as "Gold Bolt" and "Gold Mike," in keeping with a gold country theme.

Pilot James W. Carson with Errol MacBoyle's Lockheed Electra 10E.

Interior of MacBoyle's Electra 10E, like the one Amelia Earhart was flying when she disappeared in 1939. Only twenty of the planes were built.

The Loma Rica Rancho drew over 1000 visitors each year. Courteous guides were on hand to explain the history of each horse and show the awestruck visitors through the barns and grounds. Errol MacBoyle was proud of his achievements with the horse ranch, and was happy to share it with the community.

Forty-two horses were cared for at the ranch during 1940–41, and eighteen head at Inglewood. Several of MacBoyle's horses raced at Santa Anita, Tanforan, Bay Meadows and Del Mar racetracks.

Following Pearl Harbor on December 7, 1941, many changes took place in California, and one was a regulation prohibiting horse racing. This left MacBoyle with a stable full of prime two- and three-year-old horses and no place to race them.

A string of his best racehorses was transported to Omaha, Nebraska, to be entered in races in the Midwest and the East. Trainer Paul Lycan was placed in charge of the string. The decline in horse racing during the war years, and the additional costs, brought an end to MacBoyle's racing ambitions, one more disappointment linked to the war.

Loma Rica Ranch

In 1932 the Idaho Maryland Consolidated Mines Inc. purchased the 220-acre Loma Rica Ranch to prevent it from going on the auction block. The Loma Rica ranch had been planted in pears, plums, apples and cherries. Emil Larsen, from Placerville, was hired to manage the ranch.

In 1936 Errol MacBoyle purchased the Loma Rica Ranch (not the same as Loma Rica Rancho) from the mining company and continued to operate the orchard. During the depression years the pear market price dropped significantly and the other fruit prices also declined. MacBoyle gave much of the fruit away to the needy. A lot of people came and picked fruit for themselves.

MacBoyle not only mixed agriculture, mining, and raising horses, but he also raised cattle. He owned a herd of registered Guernsey cows that he pastured at the LeDuc ranch near the Brunswick mine. He also bred Shorthorn milking cows on the western section of his Loma Rica Rancho.

His bull "Rockwood Darlington" won awards in 1941 at the California State Fair and at the Cow Palace in San Francisco. In that same year he sold 20 of his milking Shorthorns at the California Breeders sale at the San Joaquin County fairgrounds in Stockton.

MacBoyle's Lake

Errol MacBoyle's lake was located on the upper elevation of the Loma Rico Rancho. The setting was so beautiful that he decided to turn the reservoir and its surroundings into a park. On the west end of the lake a bridge gave access to an octagonal gazebo and the walkway surrounding it. Large windows in each side of the gazebo provided a grand view in all directions. All of the concrete below the bridge and gazebo was faced off with rock. The outlet valves that furnished water to the mines and the ranch were located beneath the gazebo.

In 1939 MacBoyle visited the Golden Gate International Exposition (also known as the World's Fair) on Treasure Island in San Francisco Bay. There he was impressed with the "Fountain of Western Waters." He decided to have a replica built and integrated with his reservoir system. The completed fountain was located at the east end of the lake. It was outfitted with the same electrical wiring, electrical panels and lights that had been used on Treasure Island. A speaker system was also installed with the fountain, and controls for the fountain, lights and music were activated in the gazebo.

Water from the fountain flowed over two inlet waterfalls and over two fish ladders before entering the lake. Errol MacBoyle spent much of his leisure time in the gazebo, gazing at the splashing fountain and listening to music.

Grass Valley Memorial Hospital

In 1936 MacBoyle formed the Grass Valley Memorial Hospital Association. His co-directors were Edgar T. Zook, Albert Crase, Granville S. Borden and Dr. Carl P. Jones, and MacBoyle was chairman. He transferred 500,000 shares of his mine stock to the association to build and perpetuate the hospital for the community. This was to be one of MacBoyle's greatest philanthropies.

The hospital was built on the crest of the former 165-acre Prisk orchard. The building was constructed of all steel and was 80 percent complete when World War II began. From that point on it was difficult to get the necessary materials to finish the hospital. Then, in 1942, the mines were closed by government order, which caused the dividends on 500,000 shares of stock to cease, thus stopping hospital funding. The project was stopped after $425,000 dollars had been spent; another great disappointment for MacBoyle.

How Errol MacBoyle Helped

MacBoyle did a lot for his community. For example: The Grass Valley Softball Association needed equipment to erect six 75-foot light poles at the Memorial Park ball diamond. A call was made to the Idaho Maryland for assistance. Without hesitation, Errol MacBoyle sent a crane and four men to install the poles. In addition, he sent two electricians to install the light brackets and assist with the wiring.

During the depression years of the 1930s MacBoyle created work for as many men and youths as he could. The Loma Rica Rancho, Loma Rica Ranch, Loma Rica Airport, Loma Rica Reservoir, Grass Valley Memorial Hospital building and the Idaho Maryland and Brunswick mines all provided jobs. The mines were hiring men all through the depression years, and by 1941 there were over 1000 men working at the two mines.

Nevada City and Grass Valley did not feel the effects of the depression years, whereas the rest of the nation suffered greatly. Errol MacBoyle was an important reason why this community came through this difficult period without any problems.

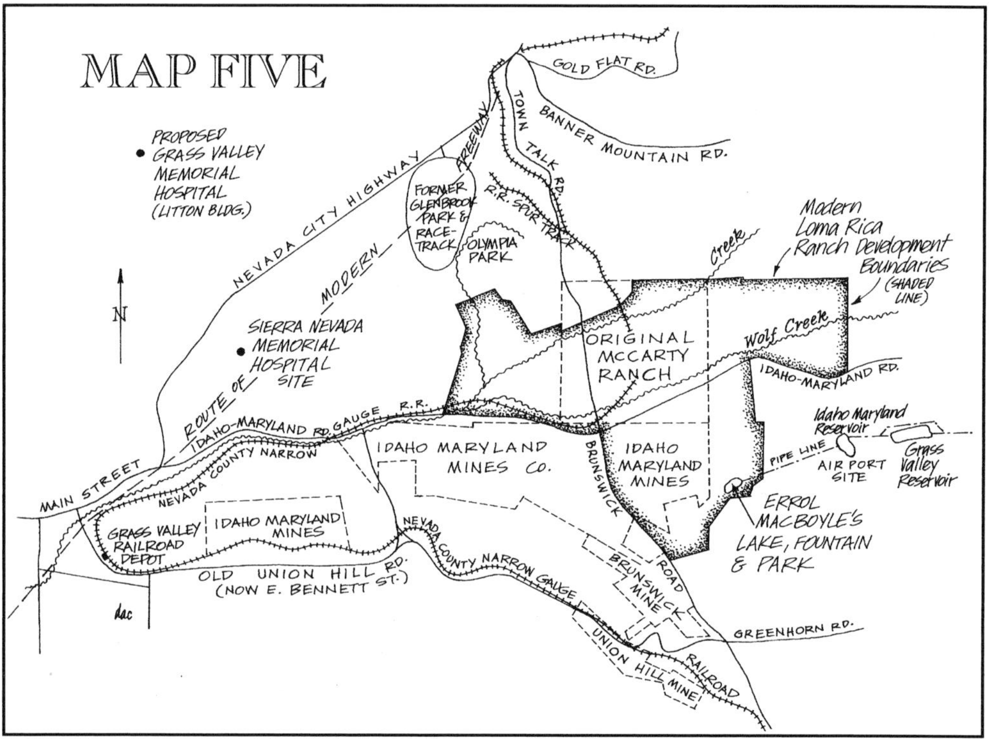

MAP FIVE

CHAPTER 13: WARTIME PROBLEMS

Magnesium was in great demand with the onset of World War II, therefore management began looking into a process for obtaining magnesium from serpentine rock. Serpentine is a silicate, and previous attempts by others to remove the magnesia contained in it had been unsuccessful, due to the fact that the serpentine resolved itself into a gelatinous mass that could not be filtered.

This difficulty was overcome by a process developed by the Idaho Maryland Mines Corp. The serpentine rock was pulverized by running it through a ball mill and then calcined under controlled temperature in an ordinary cement kiln. The calcined serpentine was agitated in a dilute solution of hydrochloric acid, thus forming magnesium chloride, which would be drawn off in liquid form through a filtering system—the residue could then be used to manufacture cement for special uses. The magnesium chloride was pure, except for a small amount of nickel and chrome. Finally, the magnesium was obtained by use of electrolytic cells to separate the nickel and chrome, which themselves were valuable byproducts.

In the spring of 1942 Twining Laboratories in Fresno, California, was used to set up a small pilot plant to prove the process was capable of producing magnesium. This process in many ways was superior to others in use, and could be produced at a lower cost. Twining Laboratories estimated the cost at less than ten cents per pound.

Errol MacBoyle went to Washington D.C. and met with various

governmental departments in an effort to interest them in the advantages of this new and successful process. He requested an examination of the pilot plant in Fresno. Months went by and no action was taken, even though private capital and idle mining equipment would be used.

The Truman committee, which had been created at the instigation of Missouri Senator Harry S. Truman, to act as the U.S. Senate's "watchdog" of the national rearmament program, elected to look into why the new magnesium process had not been considered. The committee began extensive investigations in order to confirm or disprove the representations made to the committee about the relative value of discoveries and inventions.

An examination of the pilot plant operation was made by members of the Truman committee in September 1943. They in turn requested the War Production Board to send an examiner to review the process and submit a report. The resulting WPB report was favorable in all respects. Nevertheless, further difficulty was encountered in obtaining the necessary approval for a small-scale operation to demonstrate the process further.

The Magnesium Division of the WPB explained that the nation's supply of magnesium was currently adequate. The delay in getting a new process in operation would not provide a direct contribution to the war program, they said. They also claimed the proposed operation would be in California, which had been designated as a critical manpower area. Senator Truman visited Grass Valley several times in regard to the magnesium process, and on each visit he was a guest of Errol and Glendolyn MacBoyle at their Loma Rica Rancho home.

MacBoyle's Efforts in Washington—1943

MacBoyle spent three months in Washington trying to convince the War Production Board to lift the ban on gold mining, but to no avail. He asked to at least be permitted to mine enough ore to offset the cost of pumping and timber repair. Errol MacBoyle explained to the board that their policy to transfer miners to strategic mines had failed, inasmuch as their own figures showed that fewer than 500 U.S. miners had transferred. He pointed out that most miners throughout the country had left the gold mines and secured jobs in higher-paying defense work before Order L-208 was issued.

Western gold producers and western congressmen also appealed to the WPB and the War Manpower Commission for consideration in allowing mines to reopen by employing men 60 years of age or older, and younger men who had been classified 4-F in the draft. Neither agency gave the slightest degree of consideration to the appeals.

The Idaho Maryland Mines Corp. kept close to 100 employees on the payroll between the L-208 closing date and June 28, 1943. It became necessary at that time to pare down to a bare minimum the keep the two mines pumped out and the main shafts open. The cost to keep the many miles of underground workings in good repair was too high, when no gold could be produced to offset the costs. The payroll was reduced to 60 employees.

An inspection of the underground levels of the Idaho Maryland mine was conducted eight months following the closure. In that period of time much of the timber had been weakened by decay and the swelling serpentine ground. A general failure of timber could be seen throughout the mine. Nearly all levels showed more than just minor damage.

Much of the ore in the Round Hole area, No. 201 winze levels, and the No. 3 raise area had been mined out; however, several blocks of good ore remained in place. At the current rate of timber deterioration it was highly questionable whether much of the underground workings would be accessible whenever the mine was allowed to reopen.

An inspection of the Brunswick underground levels revealed only trivial damage, although many ladderways were found to be unsafe. Unlike the Idaho Maryland mine, the Brunswick underground workings were in good hard ground, with little timbering.

On October 23, 1943, Errol MacBoyle suffered a stroke which paralyzed the right side of his body. He required constant medical care because of his illness. His stroke may have been caused in part by his efforts to reopen the gold mines and obtain approval of the magnesium process. He had spent considerable time in Washington D.C. lobbying these two causes.

Mines Reopened on a Limited Basis—1944

Through the efforts of the American Mining Congress and the San Francisco Chamber of Commerce mining committee, arrangements were made with the War Production Board to survey the damage sustained by the closure of gold mines in the Grass Valley mining district and in the Mother Lode area. The requested survey was conducted by J. R. Lane, WPB deputy director from Denver, Colorado, and A. G. Keating, WPB

These photos, taken eight months after the government-imposed L-208 closure during World War II, show major damage below the 2400-foot level—80 percent of the Idaho Maryland drifts were caved.

General manager Neil O'Donnell and Errol MacBoyle in front of MacBoyle's house.

Errol MacBoyle in front of the stables at Loma Rica Rancho in 1944.

regional mining representative. Their report of findings was forwarded to Washington for review.

Then, on May 3, 1944, after 17 months of forced closure, the Idaho Maryland Mines and the Empire Star Mines corporations were granted permission to recruit sufficient miners to produce up to 7,800 tons of ore per month, to offset the expense of keeping the mines open. The work force was limited to 200 men, who were required to be 40 years of age or older. All new men employed had to be approved by the U.S. Employment Service, and could not be essential to higher urgency war production.

The priority on obtaining critical equipment and supplies was also relaxed somewhat. These items were to be made available—however, the military and defense needs had prior call on specific items. The War Production Board agreed to assist operators in securing second-hand materials.

The fact that the underground workings had been kept free of water was a favorable factor in being able to once again produce ore, even on a limited basis. The first problem confronting the company was where to recruit qualified miners, especially because of the age restriction.

Neil O'Donnell was hired as general manager on July 1, 1944, by Errol MacBoyle, to replace Bert Crase, who had retired in the previous month. Neil O'Donnell had been in charge of the base metals section of the WPB for the importation of mercury, tin and aluminum ores. Prior to that he had managed lead and zinc mines in Bolivia. He also had a background of experience in gold mining in Idaho. Bert Crase was retained as a consulting engineer.

The Mine Workers Protective League and the mine operators agreed upon a new two-year wage and hour contract on August 9, 1944. The new contract, retroactive to July 1, provided overtime travel pay of 50 cents per day for all employees. The overtime pay was to take care of the time an employee spent traveling to and from the workplace during an eight-hour period. The agreement also carried a provision for an automatic wage increase in the event the price of gold should be raised beyond $35 per ounce.

The company was unable to recruit any additional men during the rest of the year. The total number of employees remained at 60, of which some were office workers. It was necessary to back away from some of the underground timber repair and utilize as many men as possible for mining ore, to help defray expenses.

Mining resumed exclusively in the Brunswick mine, where mill rock could be easily obtained. Recapture and timber repair continued at the Idaho Maryland, mainly to keep the 1000- and 2000-foot levels open. Production at the Brunswick mill averaged 150 tons per day. A total of 19,522 tons of ore was milled during the last six months of 1944, for a yield of $280,291. The recovery amounted to $12.14 per ton.

MacBoyle Loses Control of the Company—1945

Soon after Errol MacBoyle suffered his stroke in 1943, leaving him a semi-invalid, he heard rumors that some of the corporate directors had met to discuss selling the Idaho Maryland and Brunswick mines. At that point he realized that if he were to lose control of stock he had transferred to the Grass Valley Memorial Hospital Association, he could lose control of the mines and would be unable to prevent the sale.

In 1936 MacBoyle had transferred 500,000 shares of his own mine stock to the Grass Valley Memorial Hospital Association, which he had established to fund and perpetuate a hospital for the Grass Valley-Nevada City community. This was one of his greatest philanthropies. In addition to himself, the association board included three Idaho Maryland Mines corporate directors, namely Edgar T. Zook, Albert Crase, and Granville S. Borden (the current vice president). The other director was Dr. Carl P. Jones, from whom MacBoyle had sought guidance for the hospital layout.

Dividend income from the half a million shares was allocated 80 percent to MacBoyle until his nominal selling price of $10 per share was paid, and 20 percent to the hospital association for construction. That arrangement continued without incident for several years. By 1942, when the nation had entered World War II, the hospital was approximately 80 percent complete, at a cost of $425,000. When the mines were closed by government decree, stock dividends ceased and the association was unable to complete the hospital or to pay MacBoyle for the remaining unpaid shares of stock. The association owned, at that point, 128,000 shares, and 372,000 more had not been paid for.

Errol MacBoyle proposed the appointment of his wife Glendolyn and Dr. Robert A. Peers of Colfax to the hospital board at its meeting in September 1944. This would have increased the number on the board from five to seven members, thus assuring him control of the proxies for the 500,000 shares of stock. His suggestion was rejected by a 3 to 2 vote, with Zook, Crase and Borden voting against him. MacBoyle then asked for the resignations of the three dissenting members, but each refused.

Then Edgar Zook issued the following statement:

> "So far as the stock of the Idaho Maryland Mines Corp. owned by the hospital association is concerned, it has always been voted at all times for the best interest of the hospital, and the Idaho Maryland, and incidentally, for Mr. MacBoyle himself. It will be so voted so long as the hospital owns any stock of the Idaho Maryland Mines Corp. The majority of the board of directors of the hospital association acted for what they considered the best interests as a charitable corporation in refusing to increase the membership in the manner requested by Mr. MacBoyle."

Errol MacBoyle then brought suit against the Grass Valley Memorial Hospital Association for summary judgment for delinquent stock installments in the amount of $170,000. The case was held in Marysville on March 20, 1945, before Superior Judge Warren Steele of Yuba County. The court denied the motion for summary judgment.

Eight days later, at the annual stockholders meeting, MacBoyle, who was then the principal owner of the Idaho Maryland and Brunswick mines, suffered a major setback. He was voted out as a director at the dramatic and packed meeting. The proxies from the hospital shares were used by the three men who sat on both the hospital and the corporation boards to unseat him from control of the mines. Glendolyn MacBoyle, his wife, was elected to the board in his place, and Errol was retained as executive vice president.

Those voting to oust him were Edwin Oliver, Edgar Zook, Granville Borden, Bert Crase, and Earl Mannington. His lone supporter was Fred McNear. All had been longtime friends and associates who had benefited grandly from the faith, perseverance and inspired leadership of MacBoyle between 1926 and 1942, when he had transformed the Idaho Maryland from a "worked out" mine into the second largest gold producer in the United States.

Limitation Order L-280 Lifted—1945

In the spring of 1945, as the war in Europe was nearing an end, there was great anticipation that the ban on gold mining would be removed. Finally, on July 1, 1945, the War Production Board lifted its Order L-208. This permitted gold mining companies to resume operations, although it did not bestow any improved priority for obtaining manpower or materials.

There now was a housing shortage in Grass Valley and Nevada City caused by an influx of people working at Camp Beale and at DeWitt Hospital in Auburn. This condition had a deleterious effect on obtaining miners and other workers for mine employment.

Hiring men to work underground was gradual. Only 40 employees had been hired during the war, and by early September 1945, after Japan had surrendered, the total work force had reached just over 100 men. The lack of qualified miners and timbermen continued to plague the company. Many who applied for work did not stay long before quitting.

A small sawmill was constructed in the fall of 1945, located south of the Idaho Maryland mill, near East Bennett road. The elevated rockfill roadway entering the property from south provided a substantial dam for a log pond. The sawmill was capable of furnishing timber for mine use, and lumber for sale.

During the latter part of 1945 men began to be discharged from military service, and the cutback in defense contracts freed men to return to mining. By the end of 1945 the company was able to hire many new employees, bringing the payroll to 241 men.

Most of the men were placed in the Brunswick mine, where mill rock was readily available. Timber crews were put to work on the 1000- and 2000-foot levels in the Idaho Maryland to recapture the drifts that sustained major caving during the long shutdown. Incidentally, due to the type of swelling ground, only experienced miners and timbermen, used to that type of mining, were chosen to work in the Idaho Maryland. During 1945 there were 2,368 feet of new development headings advanced, and 543 feet of caved workings recaptured.

In 1945 there were 66,034 tons of ore produced, yielding a value of $9.77 per ton. There were 62,928 tons of ore treated in the Brunswick mill, and 3,096 tons were treated in the Idaho Maryland mill.

Errol MacBoyle Regains Control—1946

A compromise agreement was reached on January 22, 1946, between MacBoyle and the board of directors of the Grass Valley Memorial Hospital Association to end a dispute over the hospital board's influence in mine affairs. The agreement was based on the decision to operate the mine in such a manner as to make it serve the best interest of the hospital during the life of the mine.

The hospital directors granted MacBoyle a proxy on 375,000 shares of Idaho Maryland stock which he had allocated to the hospital but for which they had not yet paid. These shares, together with those he continued to hold in his own name, returned him to a position as majority stockholder and restored him to the dominant position he had held from 1926 to 1945.

The new agreement replaced the hospital board with one that would cooperate with MacBoyle and with Glendolyn MacBoyle, who was directing his affairs during his illness. The new directors were Errol MacBoyle; Neil O'Donnell, the new mine manager; Merritt Rector, retired Bank of America manager; Fred F. Cassidy, owner of Alpha Hardware; and Lynn Kelly, a Grass Valley attorney.

MacBoyle regained his seat on the corporate board of directors at the annual stockholders meeting on March 24, 1946. Earl Mannington resigned to make way for MacBoyle, but continued to serve the corporation as secretary-treasurer.

Vice president and director Edgar T. Zook retired, after handling the legal affairs of the corporation for over 21 years. William L. Oliver, of the law firm of Hall, Henry and Oliver, was elected as a director to replace Judge Zook.

The company contracted with Fred Shevlin to clean up the grounds, in and around the three mills. In addition, approximately 500 feet along Wolf Creek was dredged, and the small rock and mud was run through the old Idaho stamp mill. There was a recovery of 795 ounces of gold, with a value of over $26,000.

New Wage and Hour Agreement—1946

The Mine Workers Protective League began contract negotiations with the mine operators in April 1946 by asking for an 18½ cents-per-hour raise. The mine operators countered with an offer of 11½ cents. When a strike was threatened, the companies agreed to the union's demand. Concurrently, Empire Star Mines, Inc. suspended all mining operations. A total of 128 men at the Empire and North Star mines were terminated subsequently, leaving only 35 men to keep their four mines free of water. Most of the laid-off men were hired by the Idaho Maryland Mines Corp.

National Labor Relations Board—1946

On July 28, 1946, the Mother Lode Mine, Mill and Smelter Workers Union, Local 839, CIO, announced that the National Labor Relations Board had called upon the Idaho Maryland Mines Corp. and the union

Grass Valley Memorial Hospital under construction in the early years of World War II. Closure of the Idaho Maryland mine and the scarcity of critical building materials halted construction when 80 percent completed.

The Litton Building—intended to be Grass Valley's hospital—in 2004.

to present evidence before a NLRB hearing at the Grass Valley Veterans Memorial building.

Phyllis Lindley, secretary of Local 839, stated that the Congress of Industrial Organizations (CIO) had addressed a letter to the Idaho Maryland Mines Corp. on October 15, 1945, requesting a meeting of the company and union for the purposed of negotiating an agreement covering hours, wages and working conditions. It was further stated that the request was refused. The CIO then requested the NLRB to conduct an election among the workers at the Idaho Maryland mines to determine the collective bargaining agent for the employees.

The hearing was conducted by Robert E. Tillman, attorney for the NLRB, as trial examiner. Representing the Idaho Maryland Mines Corp. were vice president and general counsel Edgar Zook and general manager Neil O'Donnell. The Mine Workers Protective League was represented by attorney James Snell and president Russell Miller. The Mine, Mill and Smelter Workers Union was represented by CIO organizers William Gately and Phyllis Lindley. The AFL representation included state organizer Gene De Christoforo and Harry Finks, vice president of the California Federation of Labor.

Following the testimony from all parties, Tillman explained that his duty was to collect the evidence and present it to the NLRB in Washington for review and decision.

In 1937 the Twin Cities local of the Union of Mine, Mill and Smelter Workers had petitioned the NLRB regarding an action against the Idaho Maryland Mines Corp. The NLRB at that time ruled in favor of the CIO union, but that decision was overturned by the U.S. Ninth Circuit Court of Appeals, which ruled that because the corporation was not engaged in interstate commerce, the NLRB lacked jurisdiction.

Idaho Maryland Mine—1946

Due to the loss of much of the higher value ore reserves in the Idaho Maryland during the wartime closure, it now became necessary to increase development operations to find new ore bodies. Drifting on the 2700-foot level No. 2 and No. 3 veins was started as miners became available. The advancement was slow along the No. 3 vein, due to heavy ground that was influenced by the "H" and "M" faults. Recapture of the caved access drifts on the 1000- and 2000-foot levels continued at a slow and costly pace.

A new method of mining was applied to stopes in the Idaho Maryland, particularly in heavy ground. This method called for opening and stoping, from level to level, blocks of ore approximately 25 feet in length. This type eliminated a large portion of stope repair work required by the previous 100-foot-long stopes. The 25-foot blocks could be mined more rapidly, allowing that section then to be abandoned.

Brunswick Underground—1946

To compensate for the loss of the higher grade of ore from the Idaho Maryland and for the inflationary costs, it was incumbent on management to increase the tonnage of the lower grade Brunswick ore. Geologist Glenn Waterman developed a contract system for miners working in stopes, whereby they were paid extra for the amount of ore produced. The price paid varied, based upon the width of the quartz vein. This system greatly reduced the amount of waste, and cleaner ore was shipped to the mill. The result was a marked lowering of operating costs, and just as significant, the miners were happy with the increased take-home pay.

During 1946, a new method of stoping evolved. Short sublevels were driven at intervals of 20 to 30 feet along the dip of the vein. These sublevels, and ore stoped above them, were scraped to adjacent raises by slusher buckets. There were 6,552 feet of these sublevels driven during the year. This type of stope mining aided materially in maintaining a high production, while at the same time enabling management to determine the profitable and nonprofitable portions of the vein.

A high priority was given to exploring the Brunswick mine in depth, and the 3280-foot level was the logical place to begin. This provided 980 vertical feet of virgin ground below the 2300-foot level where the various Brunswick veins could be cut and a determination made of their size and values. The 3280-foot level was also within the pumping limit for raising water to the 2300-foot level with a 1,000 foot-head pump.

The shaft station, ore and waste pockets, and a sump had been completed previously. The level was established as a main haulage-way, with a 30-pound rail, a seven-foot by eight-foot cross section, and 1½-ton ore cars. The No. 2 crosscut had been previously driven 181 feet, and a priority was given to continue this crosscut in its northerly direction to cut the Brunswick and Idaho Maryland vein systems.

Sand-Slime Backfilling Stopes—1946

One of the principal improvements in the Brunswick mine during 1946 was the design and installation of equipment for filling mined-out stopes with mill tailings. Under the pressure of postwar costs, the backfilling by waste rock became increasingly expensive. The less costly sand-slime filling was perfected and used successfully. This method was not entirely new, but was modified to suit the mine conditions.

A large 24-foot by 20-foot steel tank was constructed near the mill and shaft collar, in which the tailings-sand was prepared. Measured amounts of sand and water were mixed together by means of an impeller, assisted by air jets. The pulp was held at a density of 74 percent solids and conveyed underground through a common two-inch pipe.

The stopes to be backfilled were prepared by sealing off the chutes with light timbers and caulking small openings with burlap or excelsior. Drainage of water was through a perforated close-colander wrapped in burlap, locally referred to as a "mousetrap." The tailings were distributed evenly within the stope, and in about three to four days the water was sufficiently drained to where a person could walk on the surface. The stope was filled in incremental pours.

The sandfill operation had many advantages: **a**) the solution was conveyed directly to the stope location by gravity flow; **b**) only two or three men were required for the operation; **c**) pillars left during the mining phase were able to be removed during the sandfilling phase; and **d**) this was a good means of disposing of mill tailings.

Over 5,400 tons of sand were placed underground during 1946, with a capacity of 200 tons of tailings distributed per shift. Assistant manager Roland "Dusty" Farmin and mill superintendent Richard Krebs were responsible for the design and perfecting of the sand-slime system. Mine superintendent John C. "Buck" O'Donnell was instrumental in the design of the perforated colander and stope preparations.

Production—1946

Operations during the year showed encouraging improvement over the previous year. The outlook for the future was quite promising. The number of employees increased from 241 in January to over 400 by the end of the year. During the year 185,943 tons of ore were produced. This represented 70.8 percent of the 263,768 tons of ore produced in 1941, with less than one half the work force.

There were 166,663 tons of ore treated in the Brunswick mill, and 19,310 tons in the Idaho Maryland mill. The gross value of $1,459,070, before mint charges of $5,994, left a net return of $1,453,076, or $7.81

Sand/sllime agitator next to a safety sign at collar of New Brunswick mine.

Geologist Glenn Waterman at the 2300-foot level, No. 1 drift, in the Brunswick mine in 1941.

per ton. In addition, a net income was received from the cleanup operation of the three mills by an outside contractor.

From non-mining operations a net income of $121,484 was realized. This sum represented earnings from sawmill operations, sales of old equipment, scrap, rentals, and a refund of 1944 income taxes. After the operating expenses of $1,415,661, there were administration expenses, depletion, depreciation, and patent-investigation, totaling $174,245, so that the overall operations resulted in a loss of $1,310 for 1946.

The sawmill sawed 9,015,000 board feet of lumber during the year, of which 7,706,000 bd. ft. were sold and 1,309,000 bd. ft. consumed by mining operations. Because of the earnings realized by operation of the small mill, a larger and more complete sawmill was built during the year, with a capacity of 40,000 bd. ft. per shift. This plant went into operation in mid-December.

Shortage of Miners—1947

Stoping in the Brunswick mine during 1947 continued on several of the veins from the 490-foot level down to and including the 1600-foot level. The No. 16 vein provided good ore on each level. The values were better than average, and the vein produced many nice pockets of specimen gold. There were numerous areas in the Brunswick mine where stoping operations could begin, had the manpower been available.

The lack of skilled miners during the postwar era presented a large problem for management, and seasonal work compounded this problem even more. The timber industry had begun to flourish, and many miners quit the mines during the summer months to fall timber, and then returned to the mine during the off-season. This became a great disruption to mine operations.

The No. 16 vein on the Brunswick 1300-foot level entered a large black slate formation. The vein broke up into numerous quartz stringers that ranged in size from mere seams to 1½ inches in width. Free gold could be found throughout the quartz, ranging in value from eight to ten dollars per ton.

The stoping first began as a normal shrinkage stope, with a raise and several Arizona-style chutes beneath the stope, from which ore was drawn down. The slate rock did not fragment well when blasted, and some of the rocks were too large to go through the chute openings. The motorman then had to blast the oversize rock into smaller pieces by using mud to secure a half-stick of dynamite against the rock.

To alleviate this problem, a crosscut was driven in the footwall of the stope, and a large beanhole was run up into the stope to allow the ore to free-fall onto the crosscut track. The ore was then loaded into cars by a mucking machine. Any oversize rock was easily broken by a rock hammer after the rock was dumped onto the grizzly at the 1300-foot-level ore pocket.

The development work and core drilling in the Brunswick continued during 1947 to maintain a good balance between stoping and future ore reserves. The long No. 2 crosscut on the 3280-foot level was continued without any significant ore bodies discovered.

Another Unsanctioned Strike—1947

The year 1947 had begun to look encouraging until suddenly a "wildcat" strike occurred at the New Brunswick mine on Thursday evening, May 8, 1947. A miner was discharged when he came to work on the night shift. High-grade gold was found in his underground clothes by one of the watchmen, while the clothes were hanging in the dry room. Under the direction of William Tankersly, president of the Mother Lode Mine, Mill and Smelter Workers Union, Local 839, the night shift walked off the job in protest to the firing.

At 8:00 p.m. approximately 200 miners met at the Women's Improvement Club building in Grass Valley, with Tankersly presiding. The miner accused of high-grading gave his side of the story, explaining that the gold had been planted in his clothing. The miners present at the meeting decided unanimously not to return to work unless the accused miner was reinstated. They also decided to meet the men coming off shift at the Idaho Maryland mine and encourage them to join their action.

On the following day additional miners were terminated by the company, most of whom where members of the CIO union. This precipitated the establishment of picket lines at both mines. The company countered with additional discharge notices, bringing the total to 30 miners. Eighty-eight employees crossed the lines to work without interruption throughout the labor disturbance. Four days later an additional 88 men reported for work. The walkout lasted only a short time before the rest of the men returned to work.

It was like history repeating itself, for a decade earlier, in 1937, the CIO union had called a strike against Idaho Maryland Mines Corp. in a like manner. In 1947, as in 1937, the union was recruiting new members in order to become recognized as the collective bargaining agent with the mines. They needed an incident such as that of May 8 in order to increase their membership. Local 839 had signed up 46 new members shortly after the first termination.

Following the unsuccessful strike, the Mother Lode Mine, Mill and Smelter Workers petitioned the NLRB to call an election to determine a collective bargaining agent for the employees of Idaho Maryland Mines Corp. On August 9, 1847, the NLRB dismissed the petition for lack of jurisdiction by the board, and referred back to the 1938 ruling by the Ninth Circuit Court of Appeals that the mine was not engaged in interstate commerce.

Underground Operations—1947

The current crew of 450 men was able to produce a tonnage equal to the amount that 800 miners had produced in the prewar years. Contributing to this result were the purchase of new drilling machines, longer drill steel, better mining methods, and the incentive program initiated to allow miners to increase their pay through increased output.

The underground work in the Idaho Maryland during 1947 occurred mainly on the No. 2 and No. 3 veins of the 2700-foot level. This ore was hoisted up the No. 30 winze and then trammed to the New Brunswick shaft and milled in the Brunswick mill. The recapture of caved ground in the Round Hole and No. 201 winze areas progressed very slowly.

During 1947 a total of 213,605 tons of ore was produced. The gold value was $1,748,951, with a metallurgical recovery of 93.78 percent, or $8.18 per ton. New mine exploration and development workings driven totaled 14,052 feet, and 1,573 feet of diamond-drill holes drilled.

The new sawmill proved to be very efficient, with a total of 12,201,546 board feet of timber cut during the year. The output of the mill was wholesaled to one organization only. It was necessary to erect a new sawdust burner to conform to the new state requirements.

CHAPTER 14: FIVE-DAY WORK WEEK

The California State Public Utilities Commission issued an order on March 4, 1948, for all heavy-power-consuming industries to reduce power consumption by twenty percent. An emergency power shortage had occurred because a winter drought had reduced the amount of water available to hydroelectric generating plants. In order to comply, the corporation was forced to cut back to a five-day work week at both mines and sawmills. This was the first time in the mines' history that Saturday was not a work day.

Cutting back to five days meant an automatic elimination of the time-and-a-half overtime pay. The reduced work week cut the gross payroll by nearly one third. Approximately 350 employees were affected by reduced paychecks, which in turn impacted local businesses.

A new two-year contract between the mine owners and the Mine Workers Protective League became effective on July 1, 1948. The contract granted a six cents-per-hour increase in wages, and provided for one week's vacation with pay for men employed one year or more.

Cutback in the Idaho Maryland

By mid-year only four areas were being worked in the Idaho Maryland mine. The recapture of the 1300-foot level off No. 201 winze was stopped, and an all-out effort was concentrated on mining out the No. 201 pillars from the 1300-foot level to the 1000-foot level. The last block of ore had been mined on the 900-foot level in the Round Hole area.

This ended all stoping above the 1000-foot level, leaving behind several blocks of good ore, as the cost to recapture the raises and drifts became too expensive.

The recapture work in the No. 3 raise above the 2000-foot level was advancing slowly. The raise had been reopened up close to the 1600-foot level, with plans to mine the pillars. Three raises were in progress on the 2400-foot level off the No. 45 winze. One of the raises had reached a distance of 110 feet above the level. The fourth area being worked was the No. 7 crosscut on the 2700-foot level from the No. 30 winze.

Then, in October all company work in the Idaho Maryland, with the exception of the 2700-foot level was stopped. All of the miners were moved to work in the Brunswick mine. Three leases were awarded to miners in the Idaho Maryland where the company operations left off. Edward "Ed" Wall, former mine foreman, was given the lease to remove the remaining No. 201 winze pillars and remove the rails. The Ed Wall lease also included recapturing the main No. 3 raise from the 2000-foot level and then mining the rich pillars.

Dutch Hennig was given a lease on the 2400-foot level No. 16 vein in the No. 45 winze area. This five-man crew continued stoping in good ore. The John Modrich lease, comprised of four men, began stoping in the narrow No. 16 vein on the 1500-foot level off the main Idaho shaft. The fourteen men assigned to the three leases were the only miners left working in the Idaho Maryland.

Brunswick Underground—1948

A total of over 3,500 feet of pipe and rail were removed from the No. 26 drift, No. 27 drift, and a long section of No. 1 crosscut on the 1300-foot level. Several diamond-drill holes on the No. 27 drift ran considerable water and were plugged. The No. 26 and No. 27 veins were narrow, and the grade of quartz was too low for mining.

Sending mill-tailings sand underground in a two-inch pipe continued on a regular basis. By August 1948 a total of 65,463 tons of mill tailings had been used to backfill worked-out stopes. The areas involved were on the No. 30 and No. 31 veins on the 1300-, 1450- and 1600-foot levels. Ore reserves were expanded substantially during 1948. Two outstanding developments during the year were the discovery of new ore on the 2700-foot level of the Idaho Maryland and the No. 4 vein in the Brunswick.

The long crosscut on the 3280-foot level was continued during

1948, and was stopped at 2,763 feet. The Idaho No. 3 vein and the No. 5 vein were cut, along with the No. 25 vein. The No. 3 vein was drifted upon for 403 feet, and the No. 5 vein was drifted upon for 349 feet. Neither vein appeared to have values for stoping. Each contained a serpentine footwall and a diabase hanging wall with some ankerite. Swelling ground and the need for timber prevailed, the same as the case above the 2000-foot level. The No. 25 vein was drifted upon for 193 feet. This vein produced several sample sacks of specimen gold, and stoping followed.

Employee Fatalities—1948

Two unfortunate fatalities occurred in the Brunswick mines during 1948. Leon R. Boyd was killed instantly in a blasting accident on April 8, 1948. The incident occurred in a crosscut on the 900-foot level. He had trouble lighting the round and was not far enough away from the face when the round went off.

Then on September 7, 1948, Virgil V. Storey was unaware that he was standing on a bridged area of broken rock in a raise. The rock suddenly gave way beneath him and he was quickly covered by loose rock. Storey died of suffocation before fellow workers were able to free him.

The 45 percent Trojanite dynamite used for blasting was changed to 65 percent grade. The higher-grade stick dynamite provided greater fragmentation of rock. Not only did it make ore- and waste-handling easier, but it lessened crushing in the crushing plant. Hercules powder was still used mainly for making primers in wet locations.

In order to increase efficiency in drilling operations, several vendors were invited to demonstrate their products. The Liddicoat and Oregon throw-away type bits were tested in different areas of the mine. The throw-away bits were tested against the Ingersoll jackbits that had been in use. The Oregon bit performed better than the other two. The tungsten carbide bit was also tried and found to be superior to the other three. The tungsten carbide insert was cemented in the slotted end of the drill steel. This chisel-shaped drill was found to drill faster, require less sharpening, and reduce drill breakage. This change resulted in a considerable cost savings.

There was a total of 13,847 linear feet of new development headings driven during 1948, and 7,747 feet of diamond-drill cores drilled during the year.

Production

During 1948 the company produced 192,058 tons of ore, which was 21,547 tons less than the previous year. The Idaho Maryland mine accounted for 18,444 tons, and the Brunswick mine furnished 173,614 tons of the above production. The net value of bullion delivered to the mint was $1,705,311, of which there were 45,598 ounces of gold and 13,334 ounces of silver. The value per ton was $8.88.

The mining operations showed a loss of $16,094 for the year. Depreciation, depletion and amortization of patents amounted to $163,593, making the total loss $183,688.

The sawmill operations lost money also, due to the lack of suitable logs. A total of 3,961,000 board feet of lumber was cut in 1948, down considerably from the 12,201,546 bd. ft. cut in 1947. The estimated loss in the sawmill operations, before depreciation, was $26,481.

Brunswick Underground—1949

Entering 1949 there was a great need to reduce costs. This resulted in a reduction of company personnel, and the leasing out of a large portion of the Brunswick mine. The company began removing all the broken ore in the various stopes before turning the area over to the leasing parties.

George Bird was given a lease on the No. 1 vein on the 490-foot level of the Old Brunswick mine. Sam Veale was given a lease on all levels from the 580-foot level down to and including the 1200-foot level. Dan Henry was given a lease on the levels from the 1200-foot level down to and including the 1600-foot level. The leaseholders furnished their own labor, insurance and explosives. The company provided the tools, timber, hoisting and milling. The lessees received one half of the value of gold produced in their area.

The company reserved the black slates area on the 1300-foot level, No. 16 vein, for the company account. The company also maintained all areas of the mine below the 1600-foot level, including the 2700-foot level of the Idaho Maryland. On the 3280-foot level the No. 25 drift was extended, but stopped at 261 feet, and the No. 25 drift east was stopped at 520 feet. The No. 3 drift was also extended, and stopped at 528 feet. The "H" fault intersected the No. 3 drift at 361 feet and mineralized diabase appeared at 503 feet.

The 3280-foot level was 980 feet below the nearest New Brunswick level, without any emergency exit other than by the shaft. The mine workings near the end of the long crosscut had to rely on a series of large fans and ventilation tubing for fresh air. Therefore, in order to solve these two problems, a priority was given to running a raise from the 3280-foot level to the 2700-foot level of the Idaho Maryland mine. First a crosscut had to be driven to the point where the raise was to begin. A station was cut out and a hoist set in place. A three-compartment raise was started in June 1949. A short crosscut was driven on the Idaho Maryland 2700-foot level to a location where the raise would hole through.

Development work continued on the 2700-foot level. The No. 2 drift, No. 7 crosscut, and the No. 52 drift were advanced. A large body of quartz was cut by the crosscut. A test mill run of 201 tons mined on the 2700-foot level showed $10 values. All ore and waste mined on the 2700-foot level was hoisted up the No. 30 winze and hauled to the New Brunswick shaft.

A new vein was discovered on the 2400-foot level in the No. 30 winze by diamond-drilling. The vein was quite large and was exposed for a distance of over 700 feet. The diamond-drill intercepts and associated assays of drill cores indicated that the newly discovered vein had great potential. There were insufficient development funds available to start another long crosscut to exploit the new finding. There was no available underground location close by to dispose of a large quantity of waste rock from a long crosscut. Therefore, it would be too costly to hoist the waste rock to the surface via No. 30 winze and the New Brunswick shaft.

Death of Errol MacBoyle—1949

Errol MacBoyle died in his sleep on November 4, 1949, after the long illness that had followed his stroke in 1943. The stroke had left him paralyzed on the right side of his body. MacBoyle was the one person most responsible for the Idaho Maryland Mines Corp. becoming one of the leading gold producers in the United States. Through his faith in the mines and his perseverance at times of adversity, he was able to succeed where others had failed.

He left behind a legacy for the Grass Valley-Nevada City community. "The people" were foremost in his mind. During the depression years, in addition to employing hundreds of gold miners, he formed a surface crew for men needing work, and he made sure there were jobs for students during the summer break from high school or college. MacBoyle provided employment at the Loma Rica Rancho, and created many jobs during construction of the MacBoyle airport at Loma Rica, and the Grass Valley Memorial Hospital building on top of Shaws Hill.

Production—1949

There were 7,674 linear feet of development headings driven during 1949, and 10,214 feet of diamond-drill holes drilled. Production figures for 1949 show a total of 210,152 tons of ore treated in the two mills, with 172,313 tons going to the Brunswick mill and 37,938 passing through the Idaho Maryland mill. From this tonnage 61,544 ounces of gold and 15,827 ounces of silver were produced. Metallurgical recovery was 95.6 percent of the gold. The company received $10.27 per ton milled.

Mining operations yielded earnings of $132,019 before depletion and depreciation, but after all other charges. Depletion and depreciation charges amounted to $207,269 for mining operations. A loss of $75,520 was shown after all these charges were added.

Sawmill Operations—1949

The Idaho Maryland's sawmill, capable of producing 40,000 board feet per day, was placed in operation in May 1949 under a lease agreement with Orris Donohaugh. A crew of eighteen men was employed. The 10,000 feet/per/day sawmill at the New Brunswick was leased to Mario and Lawrence Personeni.

The operations of the sawmills yielded earnings of $45,685 before depreciation, but after all other charges. Depreciation charges amounted to $31,380, thereby showing a $13,795 profit.

Brunswick Operations—1950

The wooden headframe at the Old Brunswick shaft began to weaken from dry rot and age, so replacement became necessary. A new steel headframe and an ore bin were erected. Ore hoisted through the Old Brunswick shaft from the 900-foot level continued to be trucked to the Idaho mill.

Some changes were made in the Brunswick mill to remove and treat the slime fractions separately. These included installation of a trommel screen, a classifier, and a bank of flotation cells.

The three leases in the Brunswick continued, with the George Bird lease working on the 490-foot level. Sam Veale and his crew were working from the 580-foot level down to and including the 1100-foot level. Dan Henry and his men were working from the 1100-foot level down to and including the 1600-foot level. All three leases were producing good ore.

The company stoping operation in the black slates area on the 1300-foot level continued to produce well. The large stope extended nearly to the 1200-foot level. The values in the slates remained in the narrow quartz stringers. The slate had a tendency to break into large sliver-like shapes. These pieces would fall through the grizzly at the 1300-foot level ore pocket, then cause problems during skip loading in the shaft. To alleviate this condition, a jaw crusher was installed at the grizzly, where all ore was crushed before entering the ore pocket.

The new 1880-foot level continued to be developed under company account. Crosscuts were run to explore the No. 40 and No. 16 veins. The No. 16 drift produced several sample sacks of specimen gold. Two stopes were started on the No. 16 vein. The No. 1 raise on the 3280-foot level connecting that level with the 2700-foot level of the Idaho Maryland was completed in January 1950. This raise was 350 feet in length and was important in many ways. It connected the two mines at their lowest levels, and provided a second exit in both directions.

The new raise made a big difference in air circulation in both mines. The air descending the New Brunswick shaft was increased substantially. It circulated through the 3280-foot level, then up the No. 1 raise to the 2700-foot level and up the No. 30 winze. The air then made its way through the Idaho Maryland workings to the Idaho shaft, where it was exhausted to the surface. The large ore passage in the No. 1 raise was valuable from a production standpoint. The ore from the 2700-foot level could now be hauled to the new raise and dropped down to the 3280-foot level. This eliminated hoisting up No. 30 winze.

On October 29, 1950, California Governor Earl Warren attended the centennial celebration of the 100 years of gold quartz mining in Grass Valley. Governor Warren participated in the commemoration of the Gold Hill historical marker in Grass Valley. After the activities, the governor, Congressman Harold T. "Biz" Johnson, Nevada County Supervisor Frank Rowe and Grass Valley Mayor Gilbert Cramer were given an underground tour of the Brunswick mine by mine superintendent Neil Whiting and foreman Kirk Whitmore.

Unfortunate Accident—1950

Earl Blankenship of Grass Valley was fatally injured at 8:30 a.m. on June 26, 1950, when he fell 50 feet down a raise. The accident occurred on the 800-foot level in the New Brunswick mine, where he was a member of the Sam Veale lease crew. His partner, William Tilton, was at

(Above) Foreman Kirk Whitmore, California Governor Earl Warren, Congressman Harold T. "Biz" Johnson and Grass Valley Mayor Gilbert Cramer. (Below) Lloyd "Sharkey" Roath opens the gate to let Warren and his escorts out of cage.

the foot of the raise, operating a small air-tugger hoist to lift a load of pipe up the raise. Blankenship was approximately 50 feet up in the raise, waiting to remove the pipe from the slide bucket. One of the lengths of pipe caught against a timber on the way up, causing the timber holding the cable pulley to pull loose. The timber struck Blankenship, causing him to fall to the level below. He received multiple injuries, and was pronounced dead by Dr. Vernon Padgett shortly after reaching the Jones Memorial Hospital.

Idaho Maryland Operations—1950

The company concentrated on developing the 2700-foot level during 1950. The No. 8 crosscut was driven 279 feet before being stopped. The No. 2 drift had several turns, and was not conducive as a main haulage way. The No. 9 crosscut was driven 586 feet to provide a straight haulage way from the work area. A large tonnage was anticipated, based upon the ore in sight.

Stoping was begun on the No. 2 and No. 52 veins. The ore was slushed to ore chutes by double-drum slushers. When the stope on the No. 2 vein reached the limit of the slusher, the No. 10 crosscut was driven under the stope and a raise run up into the stope. The slusher was moved closer to the face and the ore was slushed into the raise as the stope was extended.

The No. 60 winze was started in August 1950. It followed the No. 52 vein, which dipped between 15° and 20°. The vein averaged 72 inches in width, including the quartz stringers. Instead of using a conventional winze, the ore was moved by a 20-hp Ingersoll-Rand double-drum slusher. The winze advanced 146 feet by the end of the year.

Leslie and Caribeno Orzalli had just finished the No. 1 raise from the 3280-foot level to the 2700-foot level when they were assigned to run a long raise on the Idaho Maryland 2000-foot level. This raise was started on the No. 1 vein and was to connect with the 1500-foot level, where ore of a much better than average grade was mined before the war. This raise was to provide access to the No. 24 vein and some rich pillars that had not been accessible since the forced L-208 closure in 1942.

Underground Fire—1950

On Labor Day afternoon, September 4, 1950, a pumpman descended the Idaho shaft to the 2000-foot level to check the pumps. He could smell smoke, so he returned to the surface. The hoistman notified general manager Neil O'Donnell, superintendent John Charles "Buck" O'Donnell, and safety engineer Jack Clark. When they arrived at the Idaho Maryland, smoke was visible at the collar of the shaft. No one was underground because the mine was closed in observance of Labor Day.

The location of the fire was unknown, so Buck O'Donnell and Jack Clark went to the New Brunswick mine to locate the fire by following the flow of fresh air. They went to the 3280-foot level, then climbed up No. 1 raise to the 2700-foot level. They walked up the steps in No. 30 winze and found the fire just above the 2140-foot level winze station. A small cave-in of rock had broken some timbers, which in turn damaged an armored 440-volt electric cable. The arcing of the wires had ignited the nearby timber. There was very little flame, but mostly a smoldering type of burning, due to the dampness of the wood. A considerable amount of smoke resulted.

The mine rescue team was summoned and air-tight bulkheads were constructed above and below the affected area, to cut off the oxygen supply. Water was sprayed on the timber while the bulkheads were being built. After a few days the fire was out and the bulkheads removed. The underground operations at the Brunswick were not affected. The leasing operations at the Idaho Maryland were curtailed until it was certain it was safe for the men to return to work.

The Idaho Maryland and New Brunswick mines were equipped with an emergency alarm system, to be used in the event of a fire or other emergency while men were underground. This consisted of a bypass in the main air line near each shaft collar, so that methyl ethyl mercaptin could be introduced into the air supply. The liquid had a strong banana oil smell, and the resulting stench would permeate all underground work areas quickly, alerting the men of an emergency.

Corporate Changes—1950

The October 23, 1950, meeting of the board of directors of Idaho Maryland Mines Corp. at San Francisco set in play several changes in the directorate and mine management. Neil O'Donnell's resignation as executive vice president and general manager was accepted by the board. Charles Allen was chosen to succeed him as general manager, a position that now incorporated duties formerly assigned to the executive vice president, an office that no longer existed. Allen had been with the company for 15 years, serving during the past year as corporate secretary. O'Donnell remained on the board, and Herbert A. Salinger was elected as a director to replace J. D. Hoffman, who was leaving. John "Buck" O'Donnell had resigned his position as underground superintendent in September, and his was replaced by Neil Whiting, a mining engineer who had been with the company since 1939.

Corporation Losses—1950

The 1500-foot level in the Idaho Maryland mine was reopened, giving access to the rich No. 3 raise pillars and the No. 24 vein. The ore from this area provided a large part of the revenue during the latter part of the year. The corporation's deficit was reduced approximately $44,000 by operations during November and December.

During 1950 there were 193,357 tons of ore treated in the mills, resulting in recovery of 51,697 ounces of gold and 13,298 ounces of silver. The average recovery per ton of ore milled was 0.268 ounces of gold, or $9.38. The net value of this production was $1,811,669. Of this net value, the company operations produced $700,847, and the leaseholders produced $1,110,823. The lessees returned $569,087 to the company, making the total amount of company revenues $1,268,933.

The operating costs were $1,306,347. A loss of $37,414 occurred before charges for depletion and depreciation, which amounted to $179,546. The total loss for the year was $216,959 after all charges were applied. A total of 364 men were employed by the company at the end of 1950. This included surface and underground operations. In addition, 102 men were employed on the seven leases.

The company did not operate the sawmills in 1950. Both sawmills were leased on a rental basis, with $19,901.24 received from rentals. There was a loss of $27,403.77 for the year, after depreciation. The loss was largely attributable to company operation of the sawmills in 1949 and prior years.

The Need for Development Funds—1951

The year 1951 began with the agreement of a 11½ cents-per-hour raise between the Mine Workers Protective League and the owners. The wage increase and annual paid vacations affected approximately 200 Idaho Maryland personnel. Leaseholders were not included in the wage and hour agreement.

In January 1951 the corporation applied to the California Securities Commission for authority to sell 300,000 shares of treasury stock at one dollar per share par value. C. Marshall Wood, identified with a New York banking and investing business, was to purchase the shares on an option basis over a 14-month period. The Securities Commission agreed to the sale of only 200,000 shares outright to Wood after the corporation modified its original proposal.

Proceeds from the stock sales were to be used primarily for exploration and development of new ore bodies on the 1880-, 2700- and 3280-foot levels. A portion of the money was used for a study for developing and milling tungsten ores, and to pay a small bank loan.

Production was affected by a seasonal manpower drop, but experience had shown that good miners usually returned from logging or other summer employment. A number of new locations were always available where the returning men could be placed to increase output of ore. Twenty-four new lightweight jackleg drills were placed in operation, and more were to be acquired to replace older equipment as it wore out. The new jackleg drills increased footage per shift, appreciably offsetting the shortage of miners.

On January 6 and 7 a delegation of seven high-ranking Japanese mining men was given an inspection tour of the surface and underground workings of the Idaho Maryland and Brunswick mines. They were escorted by superintendent Neil Whiting and safety engineer Jack Clark. The inspection tour covered the cyanide plant, crushing plant, milling operations and specific underground operations.

These seven men were guests of the U.S. State Department and the National Resources Section of the military government under the command of Gen. Douglas MacArthur. The Japanese were studying modern mining and milling methods in order for them to be able to increase their flow of raw materials for stockpiling purposes.

Board of Directors Reorganizations—1951

Several changes in the directorate were made at the board meeting in San Francisco on March 21, 1951. Albert "Bert" Crase, who had been associated with MacBoyle between 1926 and 1944 as general manager, was elected president and general manager of the corporation. He succeeded Edwin Letts Oliver, who was retiring along with Frederick W. McNear. Both men had been associated with MacBoyle during the mines consolidation that began in 1915. They both were members of the board from 1926 to 1951. William L. Oliver, legal counsel for the board, tendered his resignation at the same meeting.

Charles Allen, who had been serving as general manager for the corporation since the resignation of Neil O'Donnell in 1950, was appointed to serve the corporation as secretary-treasurer. Members of the reorganized board of directors were Mrs. Glendolyn MacBoyle Bechhold, Bert Austin, George Harrington, Herbert H. Salinger, Max Bechhold and Bert Crase.

Underground Operations—1951

Six Brunswick hanging-wall veins appeared to dip almost uniformly toward the 3280-foot level into a large block of virgin ground. The dip of these veins indicated that the veins would converge at some lower elevation if continued. Soon after taking over management of the mines, Bert Crase ordered a crosscut run from the 3280-foot level shaft station in a southeasterly direction to explore these projected veins. A priority was given to extending the crosscut to a point far enough south where a vein could be drifted upon to come close to the No. 60 winze area of the Idaho Maryland.

The No. 60 winze continued downward a distance of 507 feet on the incline to the 2830-foot level. The winze measured 7 feet high by 18 feet wide, with an average dip of 22°. The 2760-, 2800- and 2830-foot levels were established and drifts run both east and west. Good ore was found along the entire length of the winze and in the drifts. Occurrences of high-grade ore were frequent.

The ore from sinking, drifting and stoping operations was moved by a series of double-drum slusher hoists and scraper buckets. The ore was slushed into the winze, and slushed up the winze to the 2700-foot level. The ore then was trammed to the No. 1 raise and dropped down to the 3280-foot level. A person could walk up or down the No. 60 winze and adjacent stopes with ease, thanks to the flatness of the No. 52 vein.

Stoping continued on the No. 2 vein above the 2700-foot level while drifting on the No. 51 vein was in progress. The quartz in the No. 51 vein was narrow, and was of a grade too low to mine. The No. 7, No. 11 and No. 12 crosscuts were run to explore any other veins in the general area near the No. 60 winze. Operations during 1951 were largely directed toward exploration and development of possible new ore bodies. This development work took place primarily on the 2700- and 3280-foot levels. The total amount expended in development during the year was $255,723, compared to only $105,126 in 1950. The Brunswick No. 1 crosscut on the 3280-foot level cut the No. 1 vein at a distance of 631 feet from the shaft.

Caribeno and Les Orzalli in the Idaho Maryland's No. 60 winze.

Hoistman Lauren Moody operating the double-drum slusher hoist in No. 60 winze.

Management's policy during the year was to mine the available ore of a higher value, and leave the lower grade ore. It was theorized that should the price of gold be increased, or if American gold were to be released to a free market, the corporation would be in a position to go back and mine the marginal grade of ore.

Production—1951

There were 262 men employed by the Idaho Maryland Mines Corp. at the end of the year. An average of 500 tons of ore was being milled each day, with most of it being processed in the New Brunswick mill. The Idaho Maryland mill operated on a part-time basis.

During 1951 there were 141,853 tons of ore milled, resulting in a recovery of 52,228 ounces of gold and 14,225 ounces of silver. The average recovery per ton of ore milled was $12.67. The proceeds from the gold ore amounted to $1,797,538. At the end of 1951 an estimated 70,000 tons of broken ore had accumulated in shrinkage stopes. Approximately 48,000 tons of this ore was mined in 1951.

Rentals amounting to $38,960 were received from lessees of the two sawmills owned by the corporation. After depreciation and other charges the net income was $31,384. Until now, with the exception of 1946 (the first year of operation), the sawmills had shown a consistent loss.

The Nation's Need for Tungsten—1952

When the Korean War started, the government announced a program to stimulate production of strategic and critical materials within the United States. In 1951 and 1952 the General Services Administration began a purchase program for tungsten. Producers could sell either on the open market or to the government. The purchase program was to run until 1958, or until a supply of three million units was attained, whichever came first.

Small amounts of scheelite were often found in certain quartz veins in the Brunswick mine, and in the course of mining and milling the gold ore this important metal was being lost. Management became aware that this loss could add up to several thousands of dollars each year.

Western Machinery Co. was hired to study the milling process and determine if there would be a way to save the tungsten without interrupting the gold circuit. Western Machinery's Harold Minardi recommended treating the full mill tails separately to recover the tungsten. In the meantime, the Idaho Maryland company's consulting geologist, Carlton

D. Hulin, had a different approach. He advocated that tungsten recovery could be attained by flotation treatment of the table products.

Mine geologist Charles Lyden obtained a random sample of concentrates taken from a 200-ton mill-run of gold ore from the Sam Veale lease. The sample was sent to the Carpco Engineering and Manufacturing Co. in Jacksonville, Florida, to be analyzed for scheelite. Their results showed the scheelite fraction to be 6.2 percent. Based upon the selling price of tungsten at $65 per short-ton unit of 60 percent tungsten content, an estimated loss of $1.74 per ton was indicated. This loss would not apply to all ore milled, only to certain areas where scheelite was present in a quartz vein or stringers.

Development and Stoping—1952

Development in the No. 60 winze area comprised drifting on the 2800- and 2830-foot levels during seven months of the year. The other five months were relegated to diamond-drilling several drill holes on the various winze levels. Stoping continued on the 2730-, 2860-, 2800- and 2830-foot levels.

The No. 6 crosscut on the 3280-foot level was driven in a northwest direction to a point where the No. 56 raise was to connect with the 2830-foot level in the No. 60 winze. This connection would tie the Idaho Maryland and Brunswick mines together at their lowest levels. It would provide much-needed air circulation, a second exit and less costly movement of ore.

The plan was to connect the 3280-foot level and the 2830-foot level in the No. 60 winze with a two-compartment raise. To insure that the raise would connect at the desired location, the surveying had to be perfect. A three-angle survey was started from the bottom of the No. 60 winze, then run up the winze to the 2700- foot level, then the full length of that level to No. 1 raise, down the raise to the 3280-foot level, and along the full length of that level to the No. 2 vein. The engineering and surveying over the existing survey points of both mines proved to be perfect.

The reason for this survey was to insure that the coordinates were accurate between the two mines. The No. 60 winze surveys had been taken from the Idaho shaft, whereas the 3280-foot level surveys were taken from the New Brunswick shaft. This newest long survey proved the listed coordinates and elevations of both mines were essentially accurate and matched one another.

Engineering lines were given to run the No. 8 crosscut in a position for the raise to begin. The cutout for the hoist, ore chute and raise were laid out subsequent to completion of the crosscut.

On July 30, 1952, the Idaho Maryland Mines Corp. and the Mine Workers Protective League signed a new wage agreement granting a raise of 15 cents per hour to all employees, retroactive to July 1.

All leases in the two mines were discontinued in September 1952. The company's costs had become to high to continue the lease program. Even though the leaseholders paid for their own wages, insurance and blasting powder, the company's costs for hoisting, milling, steel, timber, etc., had risen to a point where the program was unprofitable.

Revenue from mining operations and sawmill rental amounted to $1,429,937 for the year. Expenses amounted to $1,492,148, leaving an operating loss of $62,211 for 1952. The total loss for the year was $222,406, when depletion and depreciation was taken into account. There were 138,418 tons of ore milled for a return of $1,394,315. This represented a yield of 39,730 ounces of gold and 11,942 ounces of silver, for an average recovery of $10.07 per ton. During the year, 6,225 lineal feet of drifts, crosscuts and raises were driven, and 12,173 feet of diamond-drilling completed.

Grayden Beechel operating a Deitz surveying instrument underground.

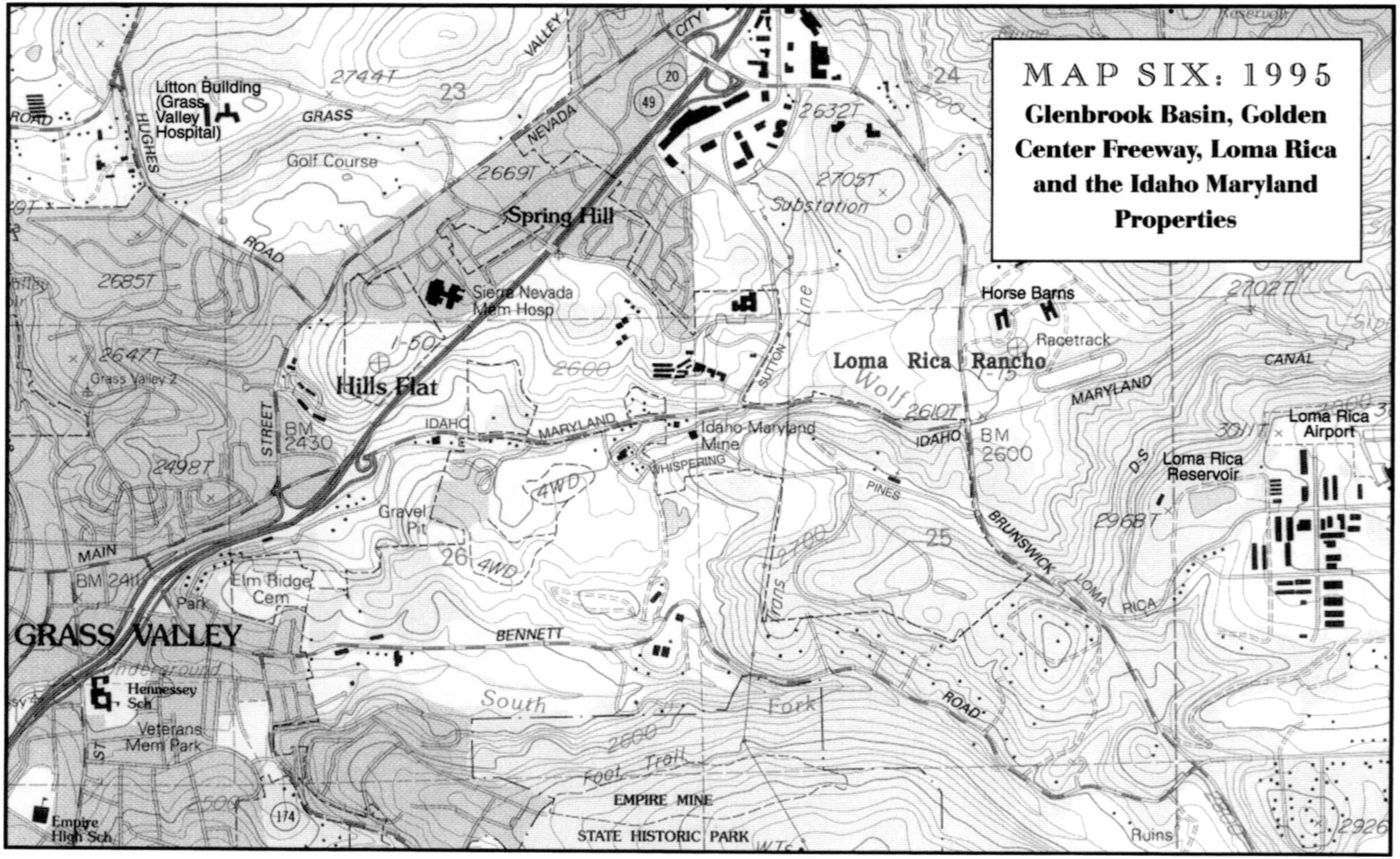

MAP SIX: 1995
Glenbrook Basin, Golden Center Freeway, Loma Rica and the Idaho Maryland Properties

CHAPTER 15: DECLINING REVENUE, HIGHER COSTS

Much of the ore tonnage in 1953 was produced in the black slates area above the 1300-foot level. The large stope on the No. 16 vein had grown to a tremendous size, measuring 40 feet wide in many places, and 140 feet long. The stope engulfed the 1200-foot level, and the values were mostly in the quartz stringers.

Other locations producing ore were the No. 25 vein on the 3280-foot level, stopes in the No. 60 winze area, and several better-grade stopes on various Brunswick levels. The mill was always hungry for ore, and management was always clamoring for higher mill heads. Diamond-drilling continued, in search of new veins, hoping to find a good grade of ore close by. Development work continued, but due to the worsening financial situation, development dollars were limited.

To conserve on the use of blasting powder, management began charging all contract miners nine dollars per box of powder. To offset this charge, each miner was allowed additional contract money in proportion to the cubic feet of rock broken in the stopes—or in the lineal feet of advance in raises, drifts or crosscuts. This proved very satisfactory in cutting costs.

Until now it had been the practice of the miners to load each drill hole clear to the collar with sticks of dynamite. After they were required to pay for their powder they asked for and used "stemming" to conserve its use. Stemming consisted of blank cartridges filled with a moist clay, and was used in place of the last two sticks of dynamite in each hole in order to tamp and seal the explosive charge. The amount of blasting powder used was reduced by 14 to 20 percent.

The mill was cut back to two shifts a day for six days a week, and it averaged 450 tons per day. The amount of concentrates was reduced proportionately to ore milled. Further savings were realized after it became necessary to close the Idaho Maryland cyanide plant and send the concentrates to the smelter at Selby.

For the first time in history, the corporation agreed to a closed-shop clause in its latest contract with the Mine Workers Protective League. Now all employees were obligated to belong to the union, with the exception of bosses, engineers and office workers. In addition, Labor Day was added as a paid holiday, and management was required to pay for a meal for employees called in to work overtime in excess of four hours.

During 1953 a greater effort was directed at finding scheelite in amounts sufficient to warrant installation of a pilot plant to test a process for full-scale milling. Exploration on the 900-foot level had revealed scheelite on the No. 19 and No. 45 veins.

More Changes in Management—1953

President and general manager Bert Crase resigned unexpectedly at the stockholders meeting on June 26, 1953. His resignation stemmed from complaints about the company's management rendered by Henry J. Nelson, who was seeking a seat on the board of directors. Nelson was associated with the investment firm of Hooke and Fay. Bert C. Austin, a widely-known mining engineer who had served on the board for two years, was elected president and general manager to succeed Crase, and Nelson was elected to take Crase's seat on the board of directors.

The Idaho Maryland Mines Corp. joined other gold-mining companies in a request for remuneration by the government for damages sustained as a result of the forced mine closures from 1942 to 1945. Sixteen of the nation's largest gold producers attended a hearing conducted by Commissioner W. E. Day in Washington D.C. before the U.S. Court of Claims. The Idaho Maryland Mines Corp. was asking for $5,021,799.

It was apparent by October 1953 that current operating losses were depleting the working capital to a critical point. For some time the operating staff had carried out a careful study of all operations, and in November recommendations were submitted that resulted in a reduction of surface and underground employees from 150 to 90. This meant a corresponding decrease in the consumption of power, supplies, payroll and general expense, without materially reducing ore production or development. The overall result of this rather drastic revision of operations showed a reduction in cost from $10.77 per ton in 1952 to $7.11 per ton in 1953.

Several parcels of land in the company's extensive holdings, not considered important to mining operations, were sold. The Idaho Maryland sawmill building was sold, but not the land it occupied. The corporation failed to pay county taxes for the 1953-1954 fiscal year because of the imperative necessity of conserving all cash resources. It was hoped that future operations would improve the situation and allow the company to catch up on its obligations.

In 1953 there were 153,634 tons of ore milled. This tonnage produced bullion and concentrates valued at $1,147,990. Mine development totaled 12,411 feet, including 6,742 feet of diamond-drilling. Ore reserves as of December 31 were estimated to be 50,973 tons of broken ore in the black slates area, with an estimated value of $8.22 per ton. There were 25,049 tons of ore in other stopes, with a value of $6.50 per ton.

Fire at the Sixteen-to-One Mine—1954

A joint mine rescue team, consisting of five members from the Idaho Maryland and five members from the Empire mine, were called to the Sixteen-to-One mine in Alleghany on January 22, 1854. The team arrived shortly before midnight and was met by state mine inspector John C. Franz and Moltier Siebenthol of the U.S. Bureau of Mines.

A fire had been discovered underground earlier in the day, and it was confined to the wooden ore bin above the Tightner shaft. A miner had noticed electrical sparking near the ore bin that was causing the fire. All of the men were evacuated from the mine soon after smoke was detected. Thick smoke caused by the smoldering damp timber, and a shortage of water in that area of the mine, greatly hampered the fire-fighting efforts. As a consequence, it took the rescue team several days to put out the fire.

Underground Operations—1954

The No. 56 raise driven from the 3280-foot level holed through as projected on the 2830- and 2800-foot levels of the No. 60 winze in January 1954. The raise continued on to where it would ultimately connect with the No. 8 crosscut on the 2700-foot level. Subsequent to the installation of grizzlies at each level, the ore from the No. 60 winze stopes was slushed to the raise and dropped down to the 3280-foot level. This greatly facilitated the movement of ore in this area of the mine. The No. 56 raise was important in that fresh-air circulation was greatly improved on the 3280-foot level, in the No. 60 winze levels and on the 2700-foot level. Previously, the air circulation had been supplied mechanically to those areas by fans and vent tubing. In addition to the reduced tram distance, a much needed second exit was provided in both directions.

Twenty-two miners were rehired in February, bringing the number of employees to 115, with mining being done in 25 locations. A priority was given to further exploration of the No. 60 winze levels, and many of these men were assigned there. The 2800- and 2830-foot levels were extended, mainly on the veins numbered 2, 52, 72 and 74. Stoping continued in this area simultaneously with the development work. There was no place to dispose of the waste rock in the No. 60 winze area, and all rock was moved as ore, causing a dilution factor that affected the mill heads.

Exploration on the 2700-foot level during the year was unsuccessful in finding any new ore. Crosscuts were driven in the footwall of the 35°-dipping stopes, and raises were run up into stopes near the face. This allowed the stopes to be extended by reducing the slushing distance. Development on the 3280-foot level was confined to the veins numbered 66, 67, 71 and 73. Stoping on that level occurred on the No. 25 and No. 63 veins.

Mining Other Minerals—1954

The exploration for scheelite in the No. 19 and No. 45 veins on the Brunswick's 900-foot level continued to find sufficient amounts to justify the installation of a milling process. In the interim, the ore was stockpiled in underground raises. George Crerar, an outstanding metallurgist from the Nevada Massachusetts mine in Imlay, Nevada, was retained as a consultant to design a flow sheet and to oversee the tungsten mill. The pilot plant was set up in the idle Idaho Maryland mill, utilizing much of the equipment used for gold processing. New equipment was added to attain the proper pulverization. Three concentrating tables were used to capture the concentrates. The concentrates were quick-roasted to magnetize the sulphides so they could be removed by a large magnet, thus eliminating the iron and other byproducts.

Approximately 100 tons of tungsten ore was processed in the experimental mill, beginning on May 14, 1954. The trial run was very successful, and the arrangement continued to be used for exploration purposes. The government purchase required 60 percent tungstic oxide (WO_3). This was easily attained, as the Brunswick scheelite ran 70 percent in its natural state. The concentrates were shipped to San Francisco in ton lots, as the company had a certificate to sell in shipments that small. The government price was $63 per twenty-pound unit.

The poor profitability of gold mining, with the low price of gold and high operating costs, was making it questionable whether gold-mining companies could continue to operate. Therefore, the board of directors decided to enter into a program of acquiring and developing uranium properties in Utah and Colorado. Here again the financial situation of the corporation limited the scope of the program, nevertheless about 3,000 acres of options, land under location, and state leases were secured on the Colorado plateau. Exploratory core drilling was carried on at several of the properties. A shipment of uranium and vanadium ore was shipped from the Flat Top Mesa holdings that netted $1,383.18.

The exploration program did not find sufficent uranium to warrant an outlay of capital to continue the search for uranium. The corporation disposed of its mining leases in Utah without any exploration. All exploratory drilling at the Colorado holdings was discontinued, and the claims on Flat Top Mesa were subleased to another company.

Financial Restraints—1954

Back in the late 1930s, when the company was very prosperous, options were entered into for the purchase of several parcels of mining properties and mineral rights. These were located for the most part south and west of the Bullion mine. As of December 31, 1953, the company had paid $151,726 on options, with a balance still owing of $178,656. The average monthly payment on these options amounted to $500, which became a financial drain on the company during the postwar years. The properties were not located adjacent to any operating mine, and to exploit the mineral rights would require new shafts, equipment and operating facilities. The corporation was under no obligation to continue the payments, which applied to the purchase price, but abandonment would mean the amounts already paid would be forfeited. The directors decided they had no choice other than to abandon all the options.

Several parcels of property owned by the corporation were sold during the year. All of the marketable timber on the mining properties was sold, bringing in $96,000. In addition to the above, the corporation was able to sell 177,500 shares of capital stock for a dollar a share at private sales. It was only as a result of these sales that the corporation was able to continue operations.

During 1954 there were 88,632 tons of ore produced, in comparison to 153,634 tons in 1953. All of the ore was hoisted through the New Brunswick shaft and milled in the Brunswick mill. The yield was 17,969 ounces of gold and 4,786 ounces of silver. Bullion with a net value of $394,886 was delivered to the mint, and concentrates with a value of $193,346 went to the American Smelting and Refining Co. in Selby, making a total amount of $558,232. In 1953 these values had been $1,088,048. The recovery in 1954 was $6.64 per ton milled, compared to $7.04 per ton in 1953.

The lower output during 1954 was partially due to the reduction of the work force from 193 to 114 men. Another factor was that, in several instances, miners were called upon to work on heavy caves on the 2000-foot level, west of No. 30 winze on the main haulage level.

The Corporation's Financial Worries—1955

In January 1955 preparations were underway to sink a winze on the 3280-foot level, where several sample sacks of high-grade gold had been gathered while driving the No. 25 drift. An area had been enlarged to set a hoist, and 40 feet of rope-raise run up above the level. Ore and waste bins were completed, and the actual sinking operations were about to begin when the operation was stopped.

The development money was limited, going into the year 1955. Aside from the tungsten exploration, all development work was run in the lower levels of both mines. Beginning in January, a drift was driven 250 feet on the No. 16 vein (more often referred to as the Morehouse vein). This drift was located on the 2000-foot level of the Idaho Maryland, close to the No. 45 winze. Two raises were also run up on the No. 16 vein. The other development headings were on the 2800- and 2830-foot levels off the No. 60 winze. The No. 1 winze on the 2830-foot level was sunk on the No. 52 vein for a distance of 70 feet before it was stopped.

A drastic cutback in employees was made in February 1955, when 48 personnel were terminated. This left a total of only 60 employees, including both surface and underground workers. By the early part of May all development work was stopped, with the exception of work involving the tungsten exploration. Mining turned to stoping whatever higher grade quartz was available.

Tungsten Mining—1955

The strong showing of scheelite on the Brunswick's 900-foot level, along with the high percentage of tungstic oxide processed in the pilot plant, helped the corporation obtain a government grant for exploration of tungsten. An exploratory project contract was procured in January 1955 from the Defense Minerals Exploration Administration. Under terms of the contract, the government agreed to reimburse the Idaho Maryland Mines Corp. for 75 percent of all expenses incurred in the exploration of tungsten. The government's participation was limited to $50,885, repayable by the corporation solely out of the gross proceeds from production of tungsten, at varying rates (limited to five percent of such proceeds).

During the period of January 20 to September 30, 1955, a significant body of scheelite ore was discovered, containing about 9,000 units of tungstic oxide. This ore shoot marked the first time in the history of the Brunswick mine that a tungsten ore body had been discovered, explored and specifically set up as a unit reserve for economic extraction of scheelite.

Previous disclosures of scheelite had been made in the gold-quartz veins of the Brunswick mine. Some attempts had been made in the past to prove out tungsten ore to justify a tungsten program. Great amounts of tungsten were allowed to be lost during milling of gold ore. This loss was at least partially justified by the requirements of gold production and the lack of proper concentration equipment needed to recover the tungsten economically.

The initial tungsten findings on the 900-foot level were in the No. 19 and No. 45 veins, but crosscutting southward from the No. 45 vein discovered the No. 46 vein. This vein, at its initial penetration, showed a body of quartz and scheelite with an average thickness of seven inches. Approximately 25 percent of the mass was scheelite. Quartz stringers were abundant in the face on both sides of the vein, showing scatterings of scheelite. The face increased in scheelite amounts as the drift was advanced.

Upon entering a scheelite-bearing drift after the miner had scaled and thoroughly wet down the vein, a person was in for a treat. When all cap lamps were turned off and an ultraviolet light turned on, a magnificent sight appeared: the fluorescent mineral and the entire last 20 feet or more of the drift seemed to come alive. Every little speck of scheelite in the muck pile, back, walls and in the face came into full glow. It was truly a sight to see.

A considerable block of high-grade tungsten ore was developed, and full-scale production followed. The pilot plant, which had been in use for exploration purposes, was no longer capable of handling the fast-increasing ore production. Arrangements were made to have the ore processed at a mill in Lovelock, Nevada. The first shipment was made on August 7, 1955. The cost of shipping and custom milling was $10 per ton. The shipping was set up on a weekly basis in special Wells Cargo trucks and trailers with a 20- to 22-ton capacity. A special loading ramp was set up at the mine to facilitate loading, and as many as 12 trucks would line up at a time to receive ore. More than 22,000 pounds of tungsten concentrate were delivered to the government stockpile in San Francisco within two months of custom milling.

The cost to ship and have the ore milled in Lovelock spurred considerable interest in planning to build a new mill for processing tungsten. The new facility would be housed in the New Brunswick mill building, utilizing much of the equipment used in the pilot plant and part of the gold circuit. George Crerar was retained again as a metallurgy consultant for the flow sheet, and he would oversee the new tungsten milling process.

On November 1, 1955, the corporation entered into a contract with Edwin Letts Oliver Inc. (Edwinco). Under terms of the contract Idaho Maryland Mines Corp. issued a note to Edwinco for $62,500, payable no later than December 31, 1956, with interest at three percent per annum. The note was to be paid out of five percent of the gross proceeds to be received from the sale of tungsten concentrates.

During 1955, there were 1,525 feet of drifts, crosscuts and raises driven under the terms of the government exploration contract. These development headings and stopes produced 2,586 tons of 1.2 percent tungsten ore, which revenue amounted to $158,970, for an average of $61.47 per ton. As of December 31, 1955, the government's reimbursements were $28,551, and the repayment thereof to $5,579, a difference of $22,972.

The corporation received a follow-up contract from the Defense

Twelve Wells Cargo trucks line up to be loaded with tungsten ore.

Dump truck loads Wells Cargo truck for the trip to Lovelock, Nevada for custom milling of the tungsten ore.

Minerals Exploration Administration for tungsten exploration of the ore body extending from the 700-foot level down to the 1100-foot level. The tungsten ore body in the Brunswick mine was situated within close proximity to the Union Hill mine, which had been full of water since its closing in 1919. This would become a concern in the coming year.

The mining and milling of gold ore was discontinued as of December 27, 1955, and all operations turned to the production of tungsten. Only a substantial increase in the price of gold would make it profitable to return to gold mining. The cost of steel, explosives and supplies used in mining had risen to a point where gold mining was no longer profitable. Unlike other businesses, there was no way to pass the increased cost on to the consumer. The price of gold was fixed at $35 an ounce, and it was sold directly to the U.S. Mint in San Francisco.

Production of gold ore in 1955 dropped significantly to 28,905 tons per year, from 88,892 tons the year before, and the grade of ore dropped to an unprofitable average of $6.32, from the only slightly better $6.64 in 1954. Operations of mine for the year resulted in a loss of $217,987 before depletion and depreciation. Income from the sales of land, equipment and timber, less non-operating expense, amounted to $143,071, reducing the net loss before depletion and depreciation to $74,516. Depletion and depreciation was $98,013, for a net loss of $23,497.

The Year of Retrenchment—1956

The year 1956 began with a general retrenchment of underground operations, beginning in the lower levels of both mines. A small crew of men began removing all trolley motors, ore cars, mucking machines, drills, hoses, slushers, etc., from all levels below the 2000-foot level, including the 3280-foot level. All pumps were left in place, and the pumping continued. The levels connecting with the No. 45 winze in the Idaho Maryland had no connection with any other underground location, therefore the No. 45 winze was allowed to fill with water after all equipment and pumps had been removed.

Now that gold mining had ceased, the future of the mine focused entirely on the production of tungsten. The exploration continued in search of scheelite on the 900-, 1000- and 1100-foot levels. At the same time, a fair grade of strategic mineral was being mined. The tungsten mill was finally completed on March 1, 1956, resulting in a very high extraction of tungsten concentrate. This mill was set up partly in conjunction with the gold-processing equipment, and had a capacity of 75 tons in 24 hours. A savings of $15 per ton was immediately realized by the elimination of transportation and custom milling charges for shipping the ore to Lovelock, Nevada.

The greatest concern involving the tungsten operation was the close proximity of the Union Hill mine to the three levels where tungsten mining was occurring in the Brunswick mine. Maps of both mines outlined the buffer zone between them, but it was not known if surveys in the Union Hill mine were up-to-date when the mine closed in 1919. The Union Hill shaft was full of water to within 20 feet of the surface.

A decision was made to dewater the shaft and upper workings down to an elevation below the lowest tungsten workings in the Brunswick. The rationale was that it would eliminate the possibility of flooding in the Brunswick mine in the event accidental contact was made with Union Hill workings. It also would provide a greater area for exploration of scheelite, where high-grade tungsten had been mined during World War I. The two-compartment Union Hill shaft was retimbered down to the water line, and a new electric line run from the Old Brunswick mine to supply power for pumping.

No appreciable advance was made in lowering the water level after a week of pumping. The project was abandoned when it became obvious that larger pumping equipment and a greater outlay of money would be necessary to remove the water. The corporation did not have the finances to dewater the Union Hill shaft, so an application was made to the Defense Minerals Administration for a grant of $122,000 for that purpose. A second application in the amount of $27,000 was submitted to pursue additional tungsten explorations in the Brunswick mine.

On February 20, 1956, the U.S. Court of Claims rendered its decision on the question of government liability in the Idaho Maryland Mines Corp. suit for damages resulting from forced closure in World War II. The court ruled in favor of the corporation and required the government to pay "just compensation" to the Idaho Maryland mines. The government filed a motion for a rehearing that was denied on July 12. Because the court's decision did not determine the amount of "just" compensation, further hearings were required to determine an equitable amount. The government appealed the case to the U.S. Supreme Court.

On March 20, 1956, the Idaho Maryland Mines Corp. offered to give a ten-acre piece of ground to the Citizens Hospital Committee for use as a hospital site. Formerly a part of the Roanaise mining claim, the land was on the crest of a hill overlooking the Nevada County golf course on the north, the city of Grass Valley on the west, and the Idaho Maryland mine on the south. Much of the credit behind this transaction can be given to Malcolm Hammill, who was the mine's surface superintendent and also served as chairman of the Grass Valley Memorial Hospital Association. Thanks to a $200,000 gift from the late Errol MacBoyle and the land donated by the Idaho Maryland Mines Corp. and the sale of debenture notes by the Citizens Hospital Committee, the Sierra Nevada Memorial Hospital was built.

Max Bechhold, manager of the mines, resigned on June 15, 1956, to administer the estate of his newly deceased brother, Sigfried Bechhold, who had married MacBoyle's widow, Glendolyn, in 1950. Safety engineer Jack Clark was appointed superintendent in charge of operations on June 15.

On September 25, 1956, orders were received from the board of directors to cease nearly all tungsten production, abandon the Idaho shaft, and to allow both mines to fill with water, up to and including the Brunswick 1450-foot level. This action was taken to reduce the number of employees to an absolute minimum consistent with the above program. The raise on the 1100-foot level was to be continued. If the government loans were approved, then men could be hired when and where necessary.

Subsequent to the decision to allow the lower levels of both mines to fill with water, the surface plant of the Idaho Maryland mine was sold to the Oro Lumber Co. The sale included the mill, cyanide plant, headframe, hoists, compressors and several buildings. The Oro Lumber Co., (also known as GPB) operators were Robert Graham, Gladys Perkins and Ed Brunning. This same company had purchased the sawmill.

All mining had been curtailed with the exception of the single raise being run on the 1100-foot level. There were great hopes that money would be forthcoming from the government to dewater the Union Hill mine and continue additional tungsten exploration in the Brunswick mine. However, word was received in December that the requested grant applications had been turned down. In addition, the California Division of Industrial Safety ordered the corporation to cease mining in close proximity to the Union Hill mine, due to the danger of flooding the Brunswick. The Brunswick mine was forced to close, leaving only a skeleton crew of men to remove all saleable equipment from the under-

For many years after most of the buildings had been removed, this was all that was visible of the New Brunswick mine. Finally only the silo remained.

A salvage crew prepares the mine property for sale.

Cover page of sales brochure for the 1957 auction of New Brunswick mine equipment.

(Above) New Brunswick mine in March 1956. (Below) Same view in 2004.

ground levels. The total production of tungsten during 1956 amounted to 8,343 tons, from which a gross of $402,000 was realized.

All gold mining operations in the Grass Valley mining district ceased in July 1956, for the first time in over 105 years. Idaho Maryland Mines Corp. had stopped its gold-mining production on December 27, 1955, when it switched operations entirely to mining tungsten. The Empire Star Mines Co. Ltd. ended its gold production on July 5, 1956, after it received a strike notice from the Mine Workers Protective League.

A Successful Auction—1957

After the mine closed, the salvage crew continued removing equipment from underground. On March 15, 1957, the last cage of items was hoisted to the surface in the New Brunswick shaft. The electric power to the mine then was disconnected at the Brunswick substation. These two great gold producers became a casualty of the low price of gold and an inflated economy that left gold mining in its wake.

On April 30, 1957, Nevada County Tax Collector Alma Hecker and Auditor/Controller John T. "Tom" Trauner jointly announced the good news that the county of Nevada and two school districts had received a check for $102,291.98 from the Idaho Maryland Mines Corp. for payment of local taxes. That amount included $34,930.33 for the current fiscal year, and $67,361.56 for delinquent taxes and late penalties. Payment of these taxes was made possible by the sale of mining equipment owned by the mine. The Milton J. Wershow and David Weisz companies of Los Angeles had been employed to auction off all saleable equipment and buildings. Beginning on May 21, 1957, a two-day auction was held at the New Brunswick mine to liquidate over 1400 lots of equipment and structures. These involved everything from the Old Brunswick, New Brunswick, and what remained of the Idaho Maryland mines. Buyers representing mining companies from many parts of the world, cities, counties, lumber mills, and interested people came to participate. Over 1,000 reviewed the items that were neatly arranged throughout the mine yard and in buildings.

The auction was a huge success, with the bidding brisk at times. Management was quite satisfied with the outcome, especially for the prices received for items such as the Marcy 86 ball mills, hoists, headframes and compressors. President Bert C. Austin announced that the money received would satisfy all outstanding debts and leave the corporation with a surplus of cash.

REFERENCES

Bean, Edwin F. *History and Directory of Nevada County, California.* Nevada City CA: Daily Gazette, 1867.

Browne, J. Ross. *Reports of the Mineral Resources of the States and Territories West of the Rocky Mountains, 1867–1868.* Washington DC: U.S. Geological Survey.

California Division of Mines. *Annual Reports, 1886–1930.*

California Journal of Mines and Geology Quarterly Mining Reports, 1935–1937.

California State Mining Bureau. *Annual Reports.*

Clark, William B. *Gold Districts of California.* Sacramento CA: California Division of Mines Bulletin 193, 1970.

Englebright, W. F. *Summary Report of the Idaho Maryland Mine for Idaho Maryland Development Company, Grass Valley, 1911.*

Farmin, Rollin. *Swelling Ground Contrasting with Heavy Ground in Mines.* AIME Technical Publications No. 1654, February 1944.

Grant, Wilbur H. *Yearly Geological Progress Reports of the Eureka and Idaho Maryland Mines in Grass Valley, 1920–1923.*

Hamilton, Fletcher. *Mines and Mineral Resources of Nevada County.* Sacramento: California Division of Mines, 1918.

Howe, Ernest. "The Gold Ores of Grass Valley," *Economic Geology* XIX, Nov. 1924.

Johnston, W. D. *The Gold Quartz Veins of Grass Valley, California.* U.S. Geological Survey, Professional Paper 194, 1940.

Lardner, W. B., and M. J. Brock. *History of Placer and Nevada Counties, California.* Los Angeles: Historic Record Co., 1924.

Lindgren, Waldemar. *The Gold Quartz Veins of Nevada City and Grass Valley Districts, California.* Washington DC: U.S. Geological Survey 17th Annual Report, 1896.

MacBoyle, Errol. *Annual Reports, 1926–1940.* Grass Valley: Idaho Maryland Mines.

———. *Mines and Mineral Resources of Nevada County.* San Francisco: California State Mining Bureau, 1919.

———. *Monthly Reports to Edwin Letts Oliver, Corporate President, Idaho Maryland Mines Corp. 1926–1934.*

Newsome, J. Branner. "Shaft Boring Found Inexpensive and Safe," *Engineering and Mining Journal,* Sept. 1936.

Raymond, Rossiter W. *Reports of the Mineral Resources of the States and Territories West of the Rocky Mountains, 1869–1877.* Washington DC: U.S. Geological Survey.

Smeaton, Tony. Collection of Errol MacBoyle correspondence regarding Idaho Maryland Mine, Grass Valley, 1925–1926.

Tolman, C. A. *General Recommendations Regarding Underground Developments, Idaho Maryland Mine, December 1932.*

Wagner, Jack R. *Gold Mines of California: Idaho Maryland, North Star and Empire Mines.* Berkeley CA: Howell-North Books, May 1969.

Wells, Harry, L. J. Albert Wilson, H. B. Rice and Allen M. Freeman. *History of Nevada County, California.* Oakland, CA: Thompson and West, 1880.

Journals and Periodicals:

Western Mining and Industrial News, 1954–1956.

Libraries:

Doris Foley Library of Historical Research, Nevada City, California.

Searls Historical Library, Nevada City, California.

Newspapers:

San Francisco *Chronicle.*

Grass Valley *Union.*

INDEX

About the Author

Jack Clark is descended from Nevada County pioneers on both sides of his family. His mother's grandfather came across the plains with a wagon train, and his father's family arrived in 1863. Jack was born in 1920 and began work as a mucker at the Idaho Maryland mine on January 9, 1941. Two months later he was promoted to the position of First Aid Attendant, and was in charge of safety and first-aid equipment above and below ground at (simultaneously) the Idaho Maryland mine, the Old and New Brunswick mines, and the Bullion mine.

During World War II, when gold mines were closed by government order, he served in the U.S. Navy Reserve and Marine Corps as a Pharmacist Mate in the South Pacific. He returned to the mine on November 3, 1945, five months after it reopened. He worked in the New Brunswick shop for 3½ years and then was given the job of Safety Engineer. Clark's last mine position was as underground superintendent from July to November 1956, when the mine ceased operations.

No one but Jack Clark could have written this book, for no other person had his unique access to operations above and below ground. As Clark says in the Preface, "In that period I acquired considerable knowledge of the operations and underground geology. With this background and my interest in mining, I felt compelled to research the mine's history and record a factual account of this famous gold producer and the men who were an important part of it."